ANTHROPOLOGICAL PAPERS OF
THE UNIVERSITY OF ARIZONA
NUMBER 78

The Winged

An Upper Missouri River Ethno-Ornithology

Kaitlyn Chandler

Wendi Field Murray

María Nieves Zedeño

Samrat Clements

Robert James

THE UNIVERSITY OF
ARIZONA PRESS
TUCSON

The University of Arizona Press
www.uapress.arizona.edu

Printed in the United States of America

22 21 20 19 18 17   6 5 4 3 2 1

ISBN-13: 978-0-8165-3202-5 (paper)

Editing and indexing by Linda Gregonis
Editorial assistance by Anna Jansson
InDesign layout by Douglas Goewey

Library of Congress Cataloging-in-Publication Data
Names: Chandler, Kaitlyn, author. | Field Murray, Wendi,
    author. | Zedeño, María Nieves, author. | Clements,
    Samrat, author. | James, Robert (Robert B.), author.
Title: The winged : an upper Missouri River ethno-
    ornithology / Kaitlyn Chandler, Wendi Field Murray,
    María Nieves Zedeño, Samrat Clements, Robert James.
Other titles: Anthropological papers of the University of
    Arizona ; v. 78.
Description: Tucson : The University of Arizona Press,
    2017. | Series: Anthropological papers ; vol. 78
    | Includes bibliographical references and index.
Identifiers: LCCN 2016042820 | ISBN 9780816532025
    (pbk. : alk. paper)
Subjects: LCSH: Ethnoornithology—Missouri River
    Valley. | Indians of North America—Missouri
    River Valley—Folklore. | Birds—Missouri River
    Valley—Folklore.
Classification: LCC E78.M82 C47 2017 | DDC
    398.2089/97—dc23 LC record available at https://
    lccn.loc.gov/2016042820

About the Authors

KAITLYN CHANDLER is an anthropological archaeologist who received her master's degree in anthropology from the University of Arizona in 2012. She is a research associate at a museum evaluation firm in the Washington, D.C. area.

WENDI FIELD MURRAY received a PhD from the School of Anthropology at the University of Arizona in 2016. She works as a research archaeologist and collections manager at the State Historical Society of North Dakota.

MARÍA NIEVES ZEDEÑO is a research anthropologist with the Bureau of Applied Research in Anthropology and a professor in the School of Anthropology at the University of Arizona.

SAMRAT CLEMENTS received a bachelor's degree in history from the University of Arizona in 2007 and is currently building a career as singer and songwriter in Worcester, Massachusetts.

ROBERT JAMES is pursuing a bachelor's degree in Anthropology from the University of Arizona.

Cover

Blackfoot Thunder Tipi, Browning, Montana, c. 1910.
Source: Glenbow Archives NA-3324-1.

Affirmation

The images of ceremonial objects were reviewed for content by Native American elders before they were chosen to be reproduced in this book.

We dedicate this book to the memory of tribal elders and friends who gave us a glimpse of their world before passing away too soon. May their souls return to the river as the most powerful winged.

The Winged

The two-wing animals.
They are the ones that fly above us.
They are the ones that see above us.
They are the ones that see the future in front of us.

—Alex Gwinn, Mandan-Hidatsa elder

Contents

TABLE

FIGURES

COLOR PLATES

Preface

The Winged: An Upper Missouri River Ethno-Ornithology is the result of a systematic study conducted by faculty and students from the Bureau of Applied Research in Anthropology in the School of Anthropology at the University of Arizona between 2010 and 2012. The study focused on bird species of the northern Plains that Native Americans living on the headwaters and trench of the Missouri River have known and used for millennia. The framework and research design for the ethno-ornithology builds on a decade of collaborative ethnographic research projects funded by the National Park Service, the U.S. Forest Service, the Bureau of Land Management, and the Indian Land Tenure Fund.

This book is truly the product of teamwork: all the authors participated in exhaustive archival searches; Wendi Murray designed the data collection instrument and, with Samrat Clements, created a qualitative database; Kaitlyn Chandler coded the information into the database and wrote the majority of the text. María Nieves Zedeño oversaw the project and converted the original report into a monograph for publication.

The Winged is also the product of enduring and trusting relationships between ethnographers and Native American elders and cultural specialists with rights to knowledge and the authority to speak about birds. Ethnographic assessments of Native American resources in Montana's Lewis and Clark National Forest and in North Dakota's Fort Union National Historic Site, Knife River Indian Villages National Historic Site, and Theodore Roosevelt National Park provided a baseline for building the ethno-ornithology. Likewise, Wendi Field Murray's master's thesis entitled *The Gods Above Have Come: A Contemporary Study of the Eagle as a Cultural Resource,* written in 2009, aided in the refinement of a methodology specifically designed for collecting and organizing information about birds.

The most important stimulus for this study is the finding that members of the Missouri River's Native American tribes and ethnic groups including, from west to east, the Blackfoot (American and Canadian divisions), Atsina, Assiniboine, Mandan, Hidatsa and their Crow relatives, and Arikara continue to use sacred bundles in their rituals. Bundles, as the material manifestation of their worldviews, generally contain one or more bird species—often a dozen or more. With the recent repatriation of many sacred bundles, tribes are actively seeking means to replenish bundle contents and to obtain bird parts for ceremonial items. It is of utmost importance, therefore, that the significance of birds in traditional lifeways be brought to light, with the goal that Native American local knowledge about this region and its bird habitats may influence conservation efforts along the fast-changing Missouri River landscape.

Acknowledgments

This book could not have been possible without the participation of Native American elders and cultural specialists, who generously trusted us with invaluable knowledge about the society of people and birds. They accepted our invitation to come to national parks and forests and talk about the cultural significance of land and resources. They graciously opened their homes, bringing out photographs and cherished objects to educate us. Multi-year funding from the National Park Service, U.S. Forest Service, and the Indian Land Tenure Fund allowed for extensive travel across the Missouri River Basin with our Native American colleagues. We are especially grateful to Dr. Michael J. Evans, whose faith in our abilities as applied ethnographers made this study possible. Over the years, University of Arizona and Plains colleagues shared information with us and helped with finding obscure data sources. Manuscript reviewers further offered comments and suggestions that vastly improved the manuscript. Anna Jansson painstakingly copy-edited the manuscript and secured permissions to publish photographs. We acknowledge the series editor, T. J. Ferguson, who invited us to submit the manuscript to the University of Arizona Press, Linda Gregonis for her editorial and indexing skills, and Doug Goewey for his InDesign expertise. Finally, we thank our spouses and pets who provided the best of sounding boards during its preparation.

CHAPTER ONE

Introduction

Early in the year 1837, the steamboat *St. Peter's* ascended the Missouri River toward trading posts and Indian villages located in the heart of the northern Plains. Unbeknownst to its captain, the steamboat boarded a man who was infected with smallpox; the virulent disease spread from each port of call the *St. Peter's* touched on its journey (Dollar 1977:18; Hollenback 2012). Fort Clark and Fort Union trading posts were two major nodes of transmission as smallpox passed from tribe to tribe through trade, horse raiding, and travel. Francis Chardon, the bourgeois of Fort Clark, recorded in detail its devastating effect on the Mandan villages. At the peak of the pestilence and turmoil at Fort Clark, the Mandan chiefs threatened Chardon with death for bringing disease purposefully, and accused him of using powerful medicine to annihilate their people (Abel 1997:24). And then,

> [while the chiefs] were talking angrily with him, he sitting with his arms on a table between them, a Dove, being pursued by a Hawk, flew in through the open door, and sat panting and worn out on Mr. Chardon's arm for more than a minute, when it flew off. The Indians, who were quite numerous, clustered about him, and asked him what the bird came to him for? After a moment's thought, he told them that the bird had been sent by the white men, his friends, to see if it was true that the Mandans had killed him, and that it must return with the answer as soon as possible; he added he had told the Dove to say that the Mandans were his friends, and would never kill him, but would do all they could for him [Audubon 1960, II:43–45].

The moment was brief, but it reveals a singularly important insight into the role of birds in Mandan society and culture. The Missouri River tribes regard birds as messengers and spiritual helpers. The ability to communicate with birds is a sign of individual spiritual power, which in this case, may have saved Francis Chardon's life.

Hundreds of species of birds frequent the Missouri River Basin, some inhabiting the region year round and some passing through seasonally on the Central Flyway that runs along the Mississippi River system from Canada to the Gulf Coast. A "flyway" is a general flight path used by birds during periods of migration, providing access to sources of food, water, and habitats, with few geographical barriers such as mountains. The Central Flyway narrows into an hourglass shape near the Platte and Missouri river valleys, offering one of the most diverse concentrations of bird species in North America (Johnsgard 2012). Not surprisingly, the Missouri River is home to many national parks, national grasslands, and wildlife refuges that once were part of tribal aboriginal territories. This astonishing world of birds has captured the attention of visitors to the Northern Plains since the time of European arrival into the region.

Early nineteenth-century explorers took copious notes on the birds (some species of which are rare or extinct today) that contribute to the understanding of the Missouri River's natural history. Members of the Corps of Discovery, notably William Clark (Jenkinson and Ronda 2003:531) and, 25 years later, Prince Maximilian de Wied (Witte and Gallagher 2008), routinely entered bird sightings in their journals. John James Audubon studied and masterfully illustrated the birds of the Missouri River (Rhodes 2006). These early adventurers gave names to many bird species that naturalists later used to create scientific taxonomies. Lesser known are bird names unique to the language and cultural traditions of the Upper Missouri

River tribes, which ethnographers recorded throughout the twentieth century. Native speakers named and classified birds according to their color, body, behavior, and birdsong. These names often carried the key to each bird's place in the social world, which they have inhabited from the time of Creation to the present.

A SOCIETY OF PEOPLE AND BIRDS

This book employs a relational approach to document human-bird interactions. A relational approach centers on the notion that humans and nature inhabit a reciprocal world (Watts 2013). It steps beyond traditional scientific and anthropological distinctions between humans and animals to reveal the intricate and eminently social character of these interactions (Zedeño 2013). Following Irving Hallowell's (1976) original analysis of the personhood of animals and things among Algonquian speakers, Eduardo Viveiros de Castro's (2004) ethnographic work in Amazonia, Anthony Gell's *Art and Agency* (1998), and Tim Ingold's *The Perception of the Environment* (2000), among other works, this book regards birds as "nonhuman agents." In the Native American world, birds are capable of engaging humans as well as other animals and things in conversation, and of influencing human behaviors and decisions. As this book demonstrates, agency transcends human-bird reciprocation; it extends this sociality to human-object (bird parts), human-landscape (bird habitats and features associated with bird trapping), and human-ancestor (bird reincarnations of human souls) "entanglements." In an entanglement, incidental contexts, events, things, and beings may amplify or modify the fundamental interaction (Hodder 2012).

Principles of ecological and cosmological knowledge are brought into focus to highlight specific beliefs, practices, and concerns associated with individual bird species. Local knowledge refers to the information and experience a community (in this case, a Native American tribe) possesses about their immediate environment including information about plants, animals, birds, migratory patterns, seasonal changes, landscapes, and their mutual relationships. One vital aspect of local knowledge is that it represents experience acquired over multiple generations of close interaction with the local environment to fulfill physical, spiritual, and cultural needs (Berkes 1993:3; Bicker et al. 2004:xi; Davis and Wagner 2003:465; Wavey 1993:13–14; also Sillitoe et al. 2002).

Tidemann and Gosler (2010:5) point out that local ornithological knowledge, or any indigenous knowledge

for that matter, is often filtered through the lens of an outsider's perspective, which can devalue the community's expertise when contrasted to Western science (also Tidemann and Whiteside 2010:154; Gosler et al. 2010). A shroud of mystery thus veils indigenous engagements with the natural world. Spirituality and practical knowledge of nature and the cosmos are presented as strange myth. More specifically, bird knowledge transmitted through oral traditions is discounted as lore. This stance diminishes the perceived quality of local information and discourages a real understanding of how humans and birds interact within specific social and cultural systems.

In contrast, this book champions the idea that indigenous knowledge systems articulate ontology, epistemology, order, and practice (e.g., Brown and Emery 2008; Mills and Walker 2008; Zedeño 2008a, 2008b, 2009, 2013). Such an integrated, dynamic system operates at multiple temporal scales, informing dispositions, decisions, and behaviors at individual and group levels. Yet, it is also subject to culturally specific and socially sanctioned rules of knowledge transfer and therefore it should be approached from a perspective that embraces plurality (Dupré 1993; Lokensgard 2010; Hollenback 2012; Murray 2011). This articulation is manifest in folk taxonomies or, more precisely, ontological taxonomies (also called taxonomies of being) that lend social and cultural order to the natural world. Ontological taxonomies are the product of phenomenological knowledge about birds and their behaviors, as well as the interconnected nature of local ecological systems, people, non-human persons or entities, and universal forces (Berlin 1992; Fabian 1975; Black 1977; Schaeffer 1950; Zedeño 2009:407).

Local users may identify birds in a different way than a scientifically trained or foreign observer would. For example, the Western Meadowlark (Plate 46), scientifically named as *Sturnella neglecta,* is known to the Blackfoot as *matsiiki* or beautiful-whistler because it signals that everything is peaceful (Hungry-Wolf 2006:136). The Greater Prairie-chicken or *Tympanuchus cupido* (Plate 16) is named for its distinctive mating "dance," which is imitated in the Prairie-chicken Dance, or *ciyó'-wagadji'bi*, that the Assiniboine adopted from the Cree. A cursory investigation into indigenous bird names can quickly reveal that descriptive names refer to symbols and meanings of a bird's character that make it powerful, as well as biographical interactions with humans and non-human persons that make it knowledgeable. The spiritual power intrinsic in some birds may be acquired by humans through dreams or visions, formally transferred from one person to another, or tapped through the incorporation of a bird or part of that bird (feather, claw, or bone) into an object such as a pipe, flute, headdress, necklace, or bundle (e.g., Irwin 1994; Murray 2009, 2011).

In turn, taxonomic principles are embedded in origin stories, oral traditions, ceremonies, and prayers that reify reciprocal relationships among humans and nature and give purpose and significance to birds beyond their role in the physical environment. For example, in the Hidatsa story of Eagle Man, recorded by Alfred Bowers (1992:470), one of two eagle friends wished to live among humans. Eagle Man chose to be carried and delivered by a Hidatsa (Awatixa) woman. He lived a long life at the Knife River villages in what is now North Dakota. Before he returned to the eagle world he gifted the Hidatsa with the Water Buster Bundle to help the people live well. Once, his eagle friend came to take Eagle Man with him. A battle ensued, and when Eagle Man was struggling to decapitate his friend, he discovered that his spinal cord was made of juneberry wood. This Hidatsa story (and many others throughout this book) highlights the intricacy of relationships among animals, plants, and people that may be biological, biographical, political, mystical, or all of the above.

It is through this incredibly rich record of Native American culture and society, especially in the Missouri River Basin, that one learns the true meaning of ontological taxonomy, where the physical boundaries of humans and natural things blur, and where humans do not always occupy the apex in the hierarchy of social relationships (in fact, they seldom do).

BUILDING A STUDY OF BIRDS AND PEOPLE

Despite their pervasive presence, birds have received relatively little attention in the regional literature, even though they are one of the most prominent animals in Plains Indian society and culture (but see Schaeffer 1950; Hungry-Wolf 1996; Moore 1986; Murray 2009, 2011). The present study was originally designed to provide federal land managers with information pertinent to past and present cultural use and significance of birds and bird parts (feathers, talons, bones, and beaks) for Native American tribes with ties to national parks and other public lands. The study followed Miki Crespi's (2003) lead in providing national parks with applied ethnographic tools for managing traditional resources in a culturally appropriate manner. The resulting monograph is much more than a management tool; it is an anthropological inquiry into the society of people and birds as found along the Missouri River.

The Winged comes at a time when concerns are growing in regard to the potential impacts of energy development on bird habitats across the North American Plains. As wildlife management policies evolve to accommodate the changing ecosystems and environmental impacts to the Missouri River, local knowledge held in tribal communities with special ties to the Missouri River and a traditional cultural investment in the river and its birds is becoming essential to the success of conservation and mitigation efforts. Equally significant is the desire, expressed by tribal members who participated in the original ethnographic research, to document and preserve traditional knowledge about birds and their significance.

Blackfoot, Assiniboine, Hidatsa, Crow, Mandan, and Arikara people have revealed (sometimes in startling detail) that birds are central to group and individual identity. This is reified through enduring and continuing cultural practices. This ancient relationship, which dates at least to the first people to inhabit North America (e.g., Hill and Rapson 2008; Falk 2002; Krech 2009; Prummel et al. 2010), is expressed in myriad ways, from contemporary individual experiences with bird power during vision quests to the repeated performance of group ceremonies that include origin stories, songs, and dances about birds. Because of their materiality and relative permanence, archaeological features play a significant role in the validation of traditional history, local knowledge, and ethnic identity as well as in current tribal cultural revitalization efforts. Descriptions of rock art, bird-trapping pits, bird effigies, buried deposits containing bird bodies and bird parts, and artifacts made of birds recur throughout this book.

The incorporation of birds into personal and communal sacred bundles is among the oldest and most salient of Native American traditions. Bundles embody principles of native ontology as well as the means to acquire, order, and transfer knowledge that is essential to the well-being of the world. Today, as before, Native Americans consider bundles to be living entities, each with its own history and personality, each with its own power and purpose (e.g., Bowers 1992, 2004; Keane 2003; Lokensgard 2010; Lowie 1922a, 1922b; Pauketat 2012; Richert 1969; Zedeño 2008a, 2009, 2013; Wildschut 1975; Wissler 1912). Bundles are more than containers of things; they embody metaphysical principles and historical trajectories that explain *what belongs with what, what doesn't, and why*. In other words, they are material manifestations of the human propensity to arrange certain things in order to strengthen their individual power by combining their particularities into an organic entity that is more than the sum of its parts.

Most Native American individuals who participated in this study own personal or medicinal bundles, or have the right to hold ceremonial bundles that belong to the community. They have the knowledge *and* the authority to speak for the birds they handle in the course of their cultural doings. In their perspective, everything is "bundled" in the sense that nothing exists in isolation and relations trump singularities. "Bundling" is not limited to religious items. Bundling is, in indigenous terms, a practical manifestation of an organic, dynamic, socialized world. To be sure, modernity and its material trappings have settled among traditions, but these are not immune to bundling and socialization—hence the book's emphasis on relationality.

This ethno-ornithology serves to update ethnographic information that was previously recorded for certain tribes in the Missouri River region. This aspect of the study is particularly relevant because the relationship that tribes have with their natural environment has evolved as the landscapes and resources they use have undergone massive changes since the mid-twentieth century and through the present day. For example, in the 1940s one-third of the Missouri River, which flows from Montana through North and South Dakota and into Nebraska, was turned into a series of reservoir lakes after the construction of six dams (Laustrup and LeValley 1998:10). The Fort Peck Dam in Montana, the Garrison Dam, which created Lake Sakakawea in North Dakota, and the Oahe and Fort Randall dams in South Dakota are among the world's largest artificial reservoirs by volume. Twenty-one percent of the land committed to the creation of reservoirs came from tribal reservations. Currently, each of these dams and their lakes shares a border with an Indian reservation, but the U.S. Army Corps of Engineers is the managing agency.

These large dams changed local and regional ecosystems, affecting the migration patterns and habitats of bird life and, consequently, the relationships local tribes have with birds. An environmental assessment performed by the U.S. Geological Survey and the Missouri River Natural Resources Committee documented how historical changes on the Missouri River have lowered the populations of fish and bird species to the point that many have been put on watch-lists as endangered or threatened species (Laustrup and LeValley 1998:9).

These alterations to the Missouri River ecosystem have impacted the river's position as a major flyway for birds, particularly eagles and waterfowl, because the dams and lakes interfere with usual migration routes. In addition, Native Americans are voicing concerns about the recent oil boom across the Plains, particularly in North Dakota. In 2010, a Mandan elder who resides on a high bluff overlooking the Missouri River in the Fort Berthold Indian Reservation expressed to the authors her concerns about oil and gas exploration on tribal land. She thinks this development is hurting waterfowl (which reincarnate women), preventing them from returning each season to help the people:

> Right now I'm so scared of all these oil wells, because I read and I see on TV these people that can just light up the water and it burns. It's got gas in the water, and my sisters [the waterbirds] got to live in that water. For that reason I agreed to help, because I'm hoping [this project] will help do something about all this pollution in the water. Because we're losing birds left and right as it is, you know. Their populations are going down, my sisters are crying and they are wondering, "where are we going to take our young ones?"

Imagine the angst of a woman whose duty is to safekeep women's knowledge, bundles, and rites, and who senses the danger development poses to the ancestors that populate the natural world and bring order and health to her people. Birds are members of a spiritual community, barometers of a community's well-being, and a tangible connection to the spiritual world that sustains them. Birds bridge the spiritual and corporeal worlds (Murray 2009).

Despite development and environmental change, the Missouri River and its bird life continue to be important resources to the region's tribes. These resources are useful for teaching younger generations about how the natural world operates and what is the appropriate behavior for living within it (Murray et al. 2011:478). This study, therefore, gathered knowledge about bird ecology and behavior as well as perceptions of environmental impacts to birds that preoccupy the regional communities.

Several tenets guide the organization of this book. The research design derived inspiration primarily from the book entitled, *Spirits of the Air,* written by Shepard Krech III in 2009. The publication of Krech's work on Native American culture and avian life in the southern United States was a major milestone in expanding the breadth of ethno-ornithology. Departing from a structure centered on taxonomy, *Spirits of the Air* examined major themes that encompass the intimate links between humans and birds and shaped southeastern indigenous worldviews and practices. Accordingly, the authors developed a flowchart of major and minor bird themes that appeared in the regional ethnographic literature and structured interview instruments around these themes. They also employed the pile-sort method to aid tribal participants in the identification and classification of familiar birds. This is a qualitative method that uses photographs of individual birds to help participants organize their knowledge about birds and to help researchers uncover principles of organization and taxonomic structures (Bernard 2006:311).

In contrast to *Spirits of the Air,* which focuses on past human-bird interactions, this ethno-ornithology integrates past and present trajectories. Archival research lends time-depth to the book's themes, while interviews with members of the tribes who hold the right to share their wisdom about birds bring traditions into contemporary focus. Although there have been ethno-ornithologies written for particular tribes (e.g., Hungry-Wolf 2006; Murray 2009, 2011; Schaeffer 1950), this study is different in its scope because it presents relevant information at a regional level, including six ethnic groups associated with the Missouri River environs, to create a single source of ethno-ornithological information with multiple perspectives. By further establishing a connection between the past and the present, the book helps to support tribal assertions of cultural affiliation and traditional association with public and private land and resources that formerly belonged in aboriginal territories.

Following Krech's (2009) and Murray's (2009) lead, the authors conducted extensive archival research of historical, ethnographic, and archaeological sources. These furnished an initial perspective on the documented cultural significance of birds that are native to the Missouri River area. We built a hierarchy of major and minor bird themes that appeared in this literature and structured interview instruments around these themes, refining them as conversations with tribal participants advanced. Bird inventories from Theodore Roosevelt National Park, Knife River Indian Villages National Historic Site, Fort Union Trading Post National Historic Site, and Glacier National Park were compiled to generate the most comprehensive bird list possible for the ethno-ornithology. This list was cross-referenced with birds specifically identified by each tribe that accepted our invitation to participate in the study (Table 1.1).

Baseline information about birds obtained from tribal interviews that were conducted over a decade of applied ethnographic research in the region were vastly augmented with targeted interviews and the application of the pile-sort method using color bird photographs (Figure 1.1), many of which are illustrated in the plates in this book. When fully integrated, different sources of knowledge drawn on for this project demonstrate and document the significance of birds including written history, oral

Figure 1.1. Calvin Grinnell (Hidatsa) sorts bird photos into organic groups in 2012. Photo by W. F. Murray.

history, ethnography, archaeology, folklore, ethnographic interviews, material culture, and visual culture. These sources further provide a deep understanding of how birds are situated in contemporary cultural practice, and what has fostered the cultural persistence of human-bird relationships that began with the creation of the world.

ORGANIZATION OF THE BOOK

After introducing the Missouri River Basin and ethnic groups represented in this ethno-ornithology to the reader (Chapter 2), the book explores seven major themes in nine chapters. Beginning with "What Makes a Bird?" (Chapter 3), these themes detail ontological, epistemological, taxonomic, and practical realms of human-bird relationships. Stories of emergence lead the way toward an understanding of the society of birds and people (Chapter 4). These traditions largely explain why birds developed certain powers and personality traits (Chapter 5). As the bridge between the natural and spiritual worlds, the book also delves into birds' unique ability to carry messages to people (chapter 6). The next two themes unpack the materiality of birds in imagery (Chapter 7) and objects (Chapters 8 and 9), and the past and present significance of birds in the realms of daily life such as trade, ritual, politics, and war. The importance of birds in subsistence is highlighted, as well as their ultimate gift to the humans who hunt and trap them to attain life's goals and dreams (Chapter 10). The book concludes with a reflection of the future of the society of people and birds.

Table 1.1. Birds Discussed in the Book.

Common Name	Scientific Name	Color Plate Number
Ducks, Geese, and Swans	**Anatidae**	
Mallard	*Anas platyrhynchos*	1
Northern Shoveler	*Anas clypeata*	
Harlequin Duck	*Histronicus histronicus*	
Common Merganser	*Mergus merganser*	2
Red-breasted Merganser	*Mergus serrator*	
Canada Goose	*Branta canadensis*	3
Snow Goose	*Chen caerulescens*	4
Tundra Swan	*Cygnus columbianus*	5
Loons	**Gaviidae**	
Common Loon	*Gavia immer*	6
Grebes	**Podicipedidae**	
Western Grebe	*Aechmorphorus occidentalis*	7
Horned Grebe	*Podiceps auritus*	
Storks	**Ciconiidae**	
Wood Stork	*Mycteria americana*	
Cormorants	**Phalacrocoracidae**	
Double-crested Cormorant	*Phalacrocorax auritus*	
Pelicans	**Pelicanidae**	
American White Pelican	*Pelecanus erythrorhynchos*	8
Herons, Bitterns, and Allies	**Ardeidae**	
Great Blue Heron	*Ardea herodias*	
American Bittern	*Botaurus lentiginosus*	9
Rails, Gallinules, and Coots	**Rallidae**	
American Coot	*Fulica americana*	10
Cranes	**Gruidae**	
Sandhill Crane	*Grus canadensis*	11
Whooping Crane	*Grus americana*	12
Lapwings and Plovers	**Charadriidae**	
Killdeer	*Charadrius vociferous*	13
Sandpipers, Phalaropes, and Allies	**Scolopacidae**	
Wilson's Snipe	*Gallinago delicata*	14
Long-billed Curlew	*Numenius americanus*	
Gulls, Terns, and Skinners	**Laridae**	
Ring-billed Gull	*Larus delawarensis*	
Partridges, Grouse, and Turkeys	**Phasianidae**	
Ruffed Grouse	*Bonasa umbellus*	
Greater Sage Grouse	*Centrocercus urophasianus*	
Sharp-tailed Grouse	*Tympanunchus phasianellus*	15
Dusky (Richardson's) Grouse	*Dendragapus obscurus*	
Spruce (Franklin) Grouse	*Falcipennis canadensis*	

Table 1.1. *(continued)*

Common Name	Scientific Name	Color Plate Number
Greater Prairie-chicken	*Tympanunchus cupido*	16
Wild Turkey	*Meleagris gallopavo*	
New World Vultures	**Cathartidae**	
Turkey Vulture	*Cathartes aura*	
Osprey	*Pandionidae*	
Osprey	*Pandion haliaetus*	17
Hawks, Kites, Eagles, and Allies	**Accipitridae**	
Golden Eagle	*Aquila chrysaetos*	18
Bald Eagle	*Haliaeetus leucocephalus*	19
Cooper's Hawk	*Accipiter cooperii*	20
Sharp-shinned Hawk	*Accipiter striatus*	21
Red-tailed Hawk	*Buteo jamaicensis*	22
Rough-legged Hawk	*Buteo lagopus*	23
Swainson's Hawk	*Buteo swainsoni*	
Northern Harrier	*Circus cyaneus*	24
Caracaras and Falcons	**Falconidae**	
Peregrine Falcon	*Falco peregrinus*	
American Kestrel	*Falco sparverius*	25
Barn Owls	**Tytonidae**	
Barn Owl	*Tyto alba*	26
Typical Owls	**Strigidae**	
Long-eared Owl	*Asio otus*	
Burrowing Owl	*Athene cunicularia*	27
Snowy Owl	*Bubo scandiacus*	28
Great Horned Owl	*Bubo virginianus*	
Eastern Screech-owl	*Megascops asio*	29
Goatsuckers	**Caprimulgidae**	
Common Nighthawk	*Chordeiles minor*	
Kingfishers	**Alcedinidae**	
Belted Kingfisher	*Ceryle alcyon*	30
Crows and Jays	**Corvidae**	
American Crow	*Corvus brachyrhynchos*	31
Common Raven	*Corvus corax*	32
Clark's Nutcracker	*Nucifraga columbiana*	33
Blue Jay	*Cyanocitta cristata*	
Gray Jay	*Perisoreus canadensis*	
Black-billed Magpie	*Pica hudsonia*	34
Pigeons and Doves	**Columbidae**	
Rock Dove	*Columba livia*	
Mourning Dove	*Zenaida macroura*	35

continued

Table 1.1. *(continued)*

Common Name	Scientific Name	Color Plate Number
Woodpeckers and Allies	**Picidae**	
Northern Flicker	*Colaptes auratus*	36, 37
Pileated Woodpecker	*Dryocopus pileatus*	38
Red-headed Woodpecker	*Melanerpes erythrocephalus*	39
Downy Woodpecker	*Picoides pubescens*	
Hairy Woodpecker	*Picoides villosus*	
Yellow-bellied Sapsucker	*Sphyrapicus varius*	
Lories, Parakeets, Macaws, and Parrots	**Psittacidae**	
Carolina Parakeet	*Cornuropsis carolinensis*	
Tyrant Flycatchers	**Tyrannidae**	
Eastern Kingbird	*Tyrannus tyrannus*	
Western Kingbird	*Tyrannus verticalis*	
Cassin's Kingbird	*Tyrannus vociferans*	
Shrikes	**Laniidae**	
Northern Shrike	*Lanius excubitor*	
Loggerhead Shrike	*Lanius ludovicianus*	
Chickadees and Titmice	**Paridae**	
Black-capped Chickadee	*Poecile atricapillus*	40
Creepers	**Certhiidae**	
Brown Creeper	*Certhia americana*	
Thrushes	**Turdidae**	
Mountain Bluebird	*Sialia currucoides*	41
Eastern Bluebird	*Sialia sialis*	42
American Robin	*Turdus migratorius*	43
Mockingbirds and Thrashers	**Mimidae**	
Gray Catbird	*Dumetella carolinensis*	
Longspurs and Snow Buntings	**Calcariidae**	
Snow Bunting	*Plectrophenax nivalis*	44
Chestnut-collared Longspur	*Calcarius ornatus*	
Blackbirds	**Icteridae**	
Bobolink	*Dolichonyx oryzivorus*	
Red-winged Blackbird	*Agelaius phoeniceus*	
Brewer's Blackbird	*Euphagus cyanocephalus*	
Yellow-headed Blackbird	*Xanthocephalus xanthocephalus*	45
Brown-headed Cowbird	*Moluthrus ater*	
Common Grackle	*Quiscalus quiscula*	
Western Meadowlark	*Sturnella neglecta*	46
Fringilline and Cardueline Finches and Allies	**Fringillidae**	
American Goldfinch	*Carduelis tristis*	47
White-throated Sparrow	*Zonotrichia albicollis*	

The Missouri River and Its People

The Missouri is the longest river in the United States, flowing more than 2,500 miles from its source on the eastern slope of the Rockies near Three Forks, Montana, to its confluence with the Mississippi River at St. Louis, Missouri (Figure 2.1). The river is a dynamic and powerful waterway. Historically notorious for its ever-changing and unpredictable nature, the "Big Muddy" (the Missouri River's nickname due to the large amounts of sediment carried in its waters) has always been a commanding force in the region. The river travels through a diverse range of ecological and geographical environments and is home to distinct riparian habitats for plants, animals, fish, and birds in its river channels and floodplains.

The river has changed greatly since it was first described and mapped by the Corps of Discovery in the early nineteenth century (Thwaites 1905a, 1905b), but transformations can be traced back thousands of years to the last Pleistocene glaciation and its aftermath. Before the Ice Ages, the landscape of what would become the upper Missouri River Basin formed a drainage that flowed north and east toward the Hudson Bay in eastern Canada, rather than southward as it does today. During the Pleistocene, glacial lobes extended southward from Canada into the northwestern reaches of the United States. Glaciers sculpted the gently rolling hills of the plains with glacial till in the north, carving impressive cliffs or "breaks" when the rock would allow it, and steering the river southward to create valleys and plateaus as it drained into the Mississippi River (Committee on Missouri River Ecosystem Science [CMRES] 2002:21–22). Transformation continues today as the river is managed through dams and flood-control efforts.

Today, the Missouri River originates at the confluence of the Gallatin, Madison, and Jefferson rivers in the Rocky Mountains (the Three Forks region). It then flows northward to Helena, Montana, where it veers northeastward where important tributaries that originate on the Continental Divide join the main channel, notably the Milk, Marias, Judith, Teton, and Sun rivers. The river then turns east toward North Dakota where entrenched bluffs and major tributaries include the Assiniboine, White Earth, and Souris rivers on the north and the Yellowstone, Little Missouri, Knife, and Heart rivers on the south. The Missouri then takes a southeast course, winding down through South Dakota and forming the border of Nebraska and Iowa, Nebraska and Missouri, and Kansas and Missouri, gathering the waters of numerous other tributaries before emptying into the Mississippi River near the city of St. Louis. Each of these rivers, in turn, have witnessed the evolution of Plains Indian society and culture.

The wealth of natural resources in the western Missouri River Basin led the U.S. Congress to designate 149 miles of the Upper Missouri as a component of the National Wild and Scenic River System in 1976, calling it an irreplaceable legacy of the historic American west. Congress further stated that the river, with its immediate environments, possesses outstanding scenic, recreational, geological, fish and wildlife, historic, cultural, and other similar values. The Bureau of Land Management undertook the preservation of the Upper Missouri River in a free-flowing condition and protects it for the benefit of present and future generations. The Upper Missouri River Breaks National Monument, a cooperative endeavor between the National Park Service and the Bureau of Land Management, was recently established in Montana.

Three national park units in North Dakota—Fort Union Trading Post National Historic Site, Theodore Roosevelt National Park, and Knife River Indian Villages National

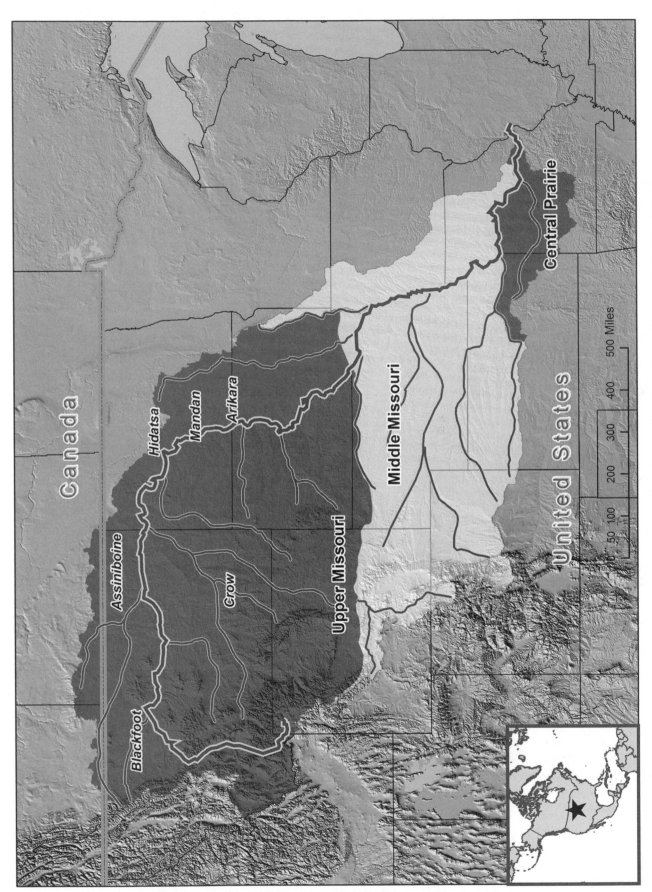

Figure 2.1. Missouri River ecoregions in relation to general location of tribes discussed in this volume. Cartography by Rita Sulkosky.

Historic Site—are located on the Missouri or on major tributaries. These parks are in the portion of the basin known as the trench. Downriver, the Lower Missouri National Recreational River and Niobrara National Scenic River straddle the South Dakota-Nebraska border. A historic-themed park unit encompassing the entire course of the Missouri River, the Lewis and Clark National Historic Trail, was created to commemorate the bicentennial of the Corps of Discovery expedition to the Pacific Ocean and back.

Large dams and reservoir lakes built in the 1940s along the Upper and Middle Missouri River had extensive and profound effects on bird habitats, populations, and consequently, the relationships that local tribes have with birds and the river. Despite these changes, the Missouri River and its associated bird life continue to be important resources to these tribes. To illustrate, Murray and colleagues (2011) explored how the Missouri River environment has been "remade" by contemporary cultures, as the artificial reservoir known as Lake Sakakawea is not only recognized as a symbol of grief and loss, but also as a place with significant connections to cultural knowledge, meaning, and continuity. Furthermore, the Missouri River as a natural and cultural resource is "used to teach younger generations about how the natural world operates and the appropriate behavior for living within it" (Murray et al. 2011:478).

Having provided a continental route to the people who entered North America more than 13,000 years ago, the Missouri River is the main vein of the Plains. Both compass and avenue, the river allowed people to move about the landscape in pursuit of bison, notable healers, trade partners, and mates, and of enemies whose scalp locks brought fame and prestige to young men (Zedeño et al. 2009). Today, U.S. Route 2, the High line, straddles the upper Missouri River, bringing people together across six Indian Reservations in Montana and North Dakota, which are home to the Blackfeet, Atsina or White Clay, Assiniboine, Ojibwa-Cree, Mandan, Hidatsa and River Crow, and Arikara or Sanish, most of whom contributed information for this book.

BLACKFOOT

The Blackfoot, or Niitsitapi, are composed of four divisions including the Blackfoot proper or Siksika, the Blood or Kainaa, the North Peigan, and the South Peigan, who are also known as the Blackfeet or Piikani.[1] All live on reserves in Canada and the United States. The Niitsitapi were bison hunters on the northwestern Plains and foothills who spoke a common Algonquian language. The landscape of the aboriginal territory, ranging from the Rocky Mountain Front across the short-grass prairie and parkland, encompasses a variety of environmental zones (Schaeffer 1950:38). The region's varying elevation and riparian environments offer a wide variety of plant and animal species and, in turn, a large range of bird species, which are prominent in Blackfoot origin traditions and ceremonial bundles.

Archaeological data, historical linguistics, population genetics, oral traditions and ritual practices support the prevailing hypothesis that the Blackfoot ancestors traversed the Rocky Mountain Front throughout the Holocene, colonizing the foothills in the last five millennia or possibly earlier (Greiser 1994; Reeves 1983, 1993; Reeves and Peacock 2001; Zedeño et al. 2014). Their aboriginal territory extended from the North Saskatchewan River to the north bank of the Yellowstone River, and from the Continental Divide to the Great Sand Hills in Saskatchewan. They maintained connections to the people and places of the Missouri River Trench through trade and visitation (Brink and Dawe 1989; Walde 2006; Zarrillo and Kooyman 2006). The Blackfeet or Piikani, in particular, had their heartland on the headwaters of the Missouri River, with their hunting ground extending to the east along its north bank and as far as the mouth of the Yellowstone River by the historic period, likely after the adoption of the horse (Jackson 2000).

The first records of the Blackfoot were written by fur traders from the competing Hudson's Bay Company and Northwest Company as they expanded their trade from eastern Canada to the North Saskatchewan River in the mid- and late-eighteenth century. Notable traders and explorers included Anthony Henday (Burpee 1907), David Thompson (Belyea 1994), Alexander Henry (Coues 1897; Gough and Brown 1988), Matthew Cocking (Burpee 1909), and Peter Fidler (Tyrrell 1934). The journals and letters left behind by these early explorers and traders establish that the Blackfoot were firmly rooted in the lands that they claim as their aboriginal territories and had been there long before European arrival.

After their first encounter with smallpox in 1780, the Blackfoot remained remote and warlike, not allowing fur traders to come and trap in their territory (Moore 2012). Beginning in 1846, when the American Fur Company established Fort Benton on the upper Missouri River at Blackfoot request, they increasingly participated in the buffalo hide trade. By the 1850s, the number of white

---

[1] Hereafter, Blackfoot refers to the language and the ethnic group in general, whereas Blackfeet refers specifically to the Blackfeet Tribe of Montana or Piikani.

settlers moving into the region began to increase and this trend continued into the 1860s with the discovery of gold. The transition to reservation life was excruciating for the Blackfoot, and tensions between them and white settlers and miners culminated with the Baker Massacre in 1870 when the U.S. Army killed 173 of their number (Ewers 1958; Jackson 2000). In the decades following the massacre, the combination of forces including decimation of bison populations, missionary reforms, a transition to ranching and farming encouraged by the Dawes Act, and cession of their Rocky Mountain lands in 1896, all chipped away at tribal lands and community life. Ceremonies survived, somehow, and so did many social and cultural aspects of Blackfoot life. Many traditions remained alive among the Canadian divisions, which were less affected by government restrictions on traditional Native culture (Bastien 2004). The Blackfeet Tribe in Montana has spent most of the twentieth century revitalizing cultural knowledge and practices with the help of Canadian elders; they recently succeeded in repatriating many sacred bundles from museums and repositories (Conaty 2015).

Relationships between the Blackfoot and birds can be traced back through the origins of the earth, animals, and people, as well as some of the most important ceremonies. The first birds were known as *Piksapi* or "mystery travelers" (North Peigan consultant). One of the first gifts from the creator Old Man Napi to the Blackfoot was to teach them how to make arrows from the split feathers of a bird in order to hunt buffalo (Grinnell 1920:140). Birds were among some of the first animals created, and they take on many roles including messengers, helpers, healers, and medicine spirits. A diverse range of birds holds significance as powerful medicine spirits, including waterbirds (loons, geese), owls, woodpeckers, corvids (magpies, ravens), raptors (hawks, eagles), and songbirds (blackbirds, chickadees), just to name a few (Zedeño et al. 2006:184–185). Their extensive ornithological vocabulary, recorded in the nineteenth century by McClintock (1999) and later by Schaeffer (1950), also reveals the importance of avian life to the group. This study adds to their linguistic and ornithological studies by including contemporary American and Canadian Blackfoot consultants.

ASSINIBOINE

The Assiniboine or *Nakoda* are a Siouan-speaking group composing one-third of the historic divisions of the Sioux linguistically close to the Dakota Sioux. A study of skeletal biology by Nancy Ossenberg suggests that, although the Assiniboine are related to the Sioux, they are not offshoots of any historic Sioux groups (Gibbon 2003:31). Biological evidence places the Assiniboine in Manitoba by A.D. 1000, while the Dakota Sioux had their base farther to the south near the Mississippi headwaters. They apparently split long before the historic period.

The movements of the ancestral Assiniboine are linked archaeologically to the mobile hunters of the parklands and boreal forests of Ontario (Zedeño et al. 2006:125). Later protohistoric and historic evidence suggests that they likely moved westward from west-central Ontario towards Lake Winnipeg in Manitoba, and then toward the lower Saskatchewan River (Denig 1961:69; Ray 1974:3–6). The turn of the nineteenth century marked a shift in the movement of the Assiniboine from a generally northwestern migration to a shift southward by 1821 (Ray 1974:94). Maximilian's 1833 account of his travels throughout the American West placed Assiniboine territory between the Missouri and the Saskatchewan rivers, bounded on the east and west by the Assiniboine and Milk rivers, respectively (Thwaites 1905a:387).

The Assiniboine, especially those living in the Woodlands, were avid participants in the fur trade as early as the establishment of the Hudson's Bay Company in 1670 (Denig 1961:69). The Plains bands of the Assiniboine contributed even more to the provisions trade that supplied trading posts with foodstuffs (Ray 1974:133–134). Fort Union was established in Assiniboine territory near the mouth of the Yellowstone River on the east side of the Missouri River in 1829 at the request of the Rock Band and after impressive returns from a winter post on the White Earth River in 1828 (Denig 1961:69–70; Larpenteur 1898:108–109, as cited in Denig 1961:70). Many traders mentioned the connection the Assiniboine had to the land around Fort Union, including accounts of a Sun Dance ceremony near the fort (Barbour 2001:131), a bison trap used by the Assiniboine only 10 miles away (Thwaites 1905a:390), and granite rock cairns placed on high bluffs with a buffalo skull for calling the buffalo (Thwaites 1905a:383). The Assiniboine were also mentioned by explorers visiting the Mandan villages along the Knife River, and were known to have been in frequent conflict with the Blackfoot, Mandan, Hidatsa, and Arikara (Denig 1961:77–79).

While the fur trade ushered in big changes for the Assiniboine, perhaps most notable were the consequences of several waves of smallpox epidemics (Ray 1974:106). After 1838, their reduced numbers could not stop the Cree from pushing their primary residences farther south, to the lands between the Cypress and Souris River on the north and the lower Milk and Missouri Rivers on the south (Ray 1974:183). Trade at Fort Union remained important to the Assiniboine, and they stayed in the area even after the fort closed in 1867.

The Fort Peck Agency was created in 1871 at an old trading post. It was later moved to a location on the Poplar River in 1880 to reduce the effects of recurrent flooding. Modern reservation boundaries for the Fort Peck Indian Reservation were established in 1886–1887 in the northwestern corner of Montana. The reservation brought together members of many different bands of Sioux and Assiniboine, including the Sisseton/Wahpetons, Yanktonais, and Teton Hunkpapa of the Sioux, and the Canoe Paddler and Red Bottom Assiniboine (Fort Peck Assiniboine & Sioux Tribes 2011). Through a different series of treaties and negotiations, however, the western bands of the Assiniboine ended up residing with the Atsina on the Fort Belknap Indian Reservation in north-central Montana. Assiniboine people also live on- and off-reserves in Saskatchewan and Alberta (Fort Belknap Indian Community 2003).

Perhaps one of the earliest descriptions of the Assiniboine concerning their relation to birds comes from an account by Jesuit Gabriel Marest in 1694, who reported that they "have large drawings on the body representing serpents, *birds*, and various other figures," which were tattooed on the skin using small, sharp bones, and wet charcoal dust (Tyrrell 1931:124, emphasis added). The most comprehensive account of the Assiniboine comes from Denig, with Lowie (1909, 1960) also contributing some minor publications on the tribe. But very little was written about Assiniboine traditional life while the elders "who remembered the buffalo culture" were still alive (Dusenberry 1960:44–45). The Montana Writer's Project (MWP) produced *Land of Nakoda*, which provides some insight into traditional resource use. Birds such as ducks and prairie-chickens were important food resources (MWP 1942:163–165). Other birds have sacred significance, which is often expressed through the use of feathers for ceremonial objects (MWP 1942:48, 96, 151). A thorough examination of historical and ethnographic resources and interviews revealed a rich history of relationships between the Assiniboine and birds, which has not received specific attention in the past, but is explored further in this book.

MANDAN, HIDATSA, AND ARIKARA

The Mandan, Hidatsa, and Arikara (MHA) Nation, also known as the Three Affiliated Tribes, have been living together for over 150 years on the Fort Berthold Reservation in North Dakota, but their individual histories go much farther back in time. The Mandan and Hidatsa are Siouan groups whose histories have been intertwined for hundreds of years, although they began as two distinct groups both linguistically and ethnically (Ahler 1986; Ahler et al. 1991; Mitchell 2013; Schneider 2001). Archaeological evidence links the Mandan and Hidatsa culture and lifestyle to the land and the Missouri River as far back as A.D. 1000, with the Mandan inhabiting the land at and above the mouth of the Heart River in the late prehistoric period (Mitchell 2013), and the Hidatsa having villages on the Knife River (Bowers 2004:8). These tribes led a semi-sedentary life based on agriculture, bison hunting, gathering and hunting the plants and animals of the Missouri River bottomlands, and living in permanent earth-lodge villages. Their livelihoods, so closely linked to the Plains landscape and the Missouri River, also formed the basis for Mandan and Hidatsa cosmology, which is entrenched in the plants, animals, birds, and features of the landscape (Beckwith 1937:xiv).

The Missouri River is central to the Mandan and Hidatsa creation stories, in which First Creator and Lone Man were once walking along on the water and decided to make land from the mud brought up from deep underwater by a diving bird (Beckwith 1937:1–3). Lone Man created land to the north, and First Creator made the land to the south, leaving a space in the middle that became the Missouri River. With the remaining mud they made Heart Butte right along the river and called it "the center of the earth" (Beckwith 1937:9, 17; Parks et al. 1978:67; Wilson 1910, as cited in Wood 1986:100). Some divisions of the Mandan people were created in the south. They travelled up the Mississippi and the Missouri rivers to eventually settle at the mouth of the Heart River (Bowers 2004:vii, 156–157). The Hidatsa likely came to the Missouri River from the northeast, after the Mandan, and were advised by the Mandan to settle "close enough to be friends and not far enough away to be enemies" (Bowers 1992:15). They travelled to the Knife River area and set up villages there. It was around that time that the Hidatsa divided, with some remaining in the region to practice agriculture while others left and became the River Crow (Bowers 1992:15–16).

The Arikara come from the Caddoan-speaking groups of the central Plains of Kansas and Nebraska, southeast of our study area (Parks 2001). In their origin story, the Arikara people emerged from underground and embarked on a journey westward, led by the sacred Mother Corn (Dorsey 1904; Gilmore 1930). They gradually moved westward and northward up the Missouri River, leaving the Pawnee sometime after the fourteenth century. The Arikara moved up through the Middle Missouri region during the late prehistoric period, arriving in the Big Bend area of the Cheyenne River sometime after A.D. 1300. By 1675 the historical Arikara tradition was present along the Missouri. After a series of smallpox epidemics and attacks from neighboring tribes, the Arikara joined the Mandan and Hidatsa at Like-a-Fishhook Village in 1862, where they banded together in an effort to preserve their cultures and people during the nineteenth century (Murray and Swenson 2016). These tribes became the Three Affiliated Tribes and have lived together in the Fort Berthold Reservation since its creation in 1870.

Epidemics, loss of land, and loss of livelihood threatened Mandan, Hidatsa, and Arikara cultural practices, such as the elaborate bundle systems that hold the history and spiritual knowledge of the tribes. Many bundles have been lost or retired over the years, especially as a result of two major smallpox epidemics that killed many bundle-holders before they had the opportunity to transmit their knowledge to the next generation. The Missouri River continued to be a central feature of Mandan, Hidatsa, and Arikara identity, despite the reductions of their traditional lands through a series of treaties, including the Treaty of Fort Laramie in 1851 and subsequent reductions of lands culminating in the creation of the Fort Berthold Reservation. Over the years since the tribes joined together, close relationships and intermarriage have created new bonds, which make it possible for oral traditions, ceremonial societies, sacred bundles, and other important traditional knowledge to be shared (Zedeño 2009:iv). Yet, Arikara identity remains separate from the Mandan and Hidatsa.

Birds play a major role in the creation story of these groups, as well as in their traditional history, beliefs, and practices. The importance of birds in these cultures may be partially attributed to the fact that the Missouri River Trench, where these tribes are settled, is a major migratory path for many species. Many birds, including eagles, play a prominent role in the traditions and hunting practices of these tribes. Major changes on the landscape, however, have altered the relationship between these people and their traditional territories.

The 1947–1953 construction of the Garrison Dam on the Missouri River near Riverdale, North Dakota, created Lake Sakakawea and radically altered the landscape and ecosystems on that stretch of the river. Further, the reservoir almost exclusively affects lands within the Fort Berthold Indian Reservation, ultimately inundating 152,360 acres, about one-fifth of the tribal lands (Lawson 1982, 1994; Murray et al. 2011:471–472). Many community members were forced to relocate. Entire towns were flooded, the best farming lands were destroyed, and culturally and historically important places on the landscape were lost

(Murray et al. 2011:473). Some bundles left behind by deceased elders before they could pass them down were left in homes flooded by the waters of Lake Sakakawea because relatives were hesitant to move these powerful entities without the proper knowledge or rights. The loss of traditional knowledge is comparable in some ways to the damage sustained during the smallpox epidemics of the eighteenth and nineteenth centuries (Hollenback 2010, 2012; Murray et al. 2011:473).

Despite the losses caused by the construction of the Garrison Dam, the Mandan, Hidatsa, and Arikara continue to find ways to connect with the changed landscape in ways that remember their past and look towards the future, recognizing that "resilience in the face of change remains an integral part of their history and identity" (Murray et al. 2011:469). The waters of Lake Sakakawea still hold associations with the Missouri River that feeds it, giving the lake a sacred significance despite its change on the landscape. One Fort Berthold consultant explained:

> That water is holy and sacred. My mom and grandma used to tell me, you want to get in the water, don't be playing in that water, it's holy stuff. It's all connected, this water. It comes from that direction. These clouds, they come from Thunderbird. He's a big eagle, a live eagle. They all are part of this. This is all God [Murray et al. 2011:476].

Changes in the landscape and ecology of the Missouri River as a result of the construction of the Garrison Dam, federal regulations on eagle trapping and possession of eagle materials, and changes in land use and subsistence have all contributed to changing relationships among the people, the land, and the birds, especially the eagle (Murray et al. 2011). Tribal members have expressed concern about the effects of the lake on the bird life in the area, particularly eagle migrations. These changes affect their spiritual health when the birds cannot return to the area to help the people.

CROW

The Crow take their name from the native word *Absarokee* or *Apsáarooke*, which roughly translates to "children of the large beaked bird," widely interpreted as a crow (Linderman 2002:28; McCleary 1997:1; Voget 1995:xvii, 2001). Plenty Coups (1848–1932), the last hereditary chief of the Crow, translated the tribe's name to mean "Children of the Raven," while Voget (1995) posits that the large beaked bird could also refer to the eagle. Eagles are known to the

Crow as "Big Bird" when they take on the role of Thunderbird (Voget 1995). Some of the first French explorers to encounter the Crow referred to them as *les beaux hommes*, or Handsome Men (Burpee 1927), and French traders later referred to them as *gens du corbeau(x)*, or People of the Crow (Voget 1995:xvii). The Crow's connection to the avian world is all but indisputable from their name alone.

Historically the Crow have been grouped into three smaller bands, the Mountain Crow (*Ashalahó*, "Where there are many lodges"), the River Crow (*Binnéessiippeele*, "Those Who Live amongst the River Banks"), and the Kicked in the Bellies (*Eelalapíio*, "Kicked in the Bellies," or *Ammitaalasshé*, "Home Away from The Center") (Lowie 1956; McCleary 1997:2). These groups live together in the Crow Indian Reservation in southern Montana near Billings, but they retain their band ties, which are reflected in residential districts throughout the reservation.

Archaeological, historical, and linguistic evidence indicates that the Crow are closely related to the Hidatsa. Scholarly consensus suggests that the Crow may have split off from the Hidatsa to become a distinct group as early the late prehistoric period, or perhaps as late as the eighteenth century (Wood and Downer 1977). Linguistic evidence suggested to Wood and Downer that a split in the Siouan-family languages of the Crow and Hidatsa occurred at least 500 years ago and possibly earlier (Reed n.d.). Archaeological evidence also links the Crow and Hidatsa through material culture and sites found along the Bighorn Mountains (Frison 1979). Crow bands may have formed from the Hidatsa as a result of series of splits and migrations. The migrations included a temporary stay among the Blackfoot (MCleary 1997:16; Peck 2011:437). The Bighorn Mountains, which crosscut the Montana-Wyoming state line, eventually became the center of the Crow world. Larocque, a French trader and explorer, gives the earliest European account of Crow territory, which ranged from the foot of the Rocky Mountains along the Yellowstone River in the Spring and Fall, to the Tongue River and Pryor Creek in the Summer (Denig 1980:21).

The introduction of the horse to the Plains greatly changed Crow mobility, settlement patterns, and bison hunting practices. Horses allowed the Crow to move their base camps more often, making chasing bison herds a more profitable endeavor. Many fur traders made mention of the Crow as horsemen and warriors, although intertribal aggression on the Missouri River prevented substantial involvement of the Crow in the fur trade until the establishment of Fort Union Trading Post by the American Fur Company in 1829 (Voget 2001; Wood and Thiessen 1985). The Fort Laramie Treaty of 1851 established a reservation for the Crow in south-central Montana and northern Wyoming. This area was greatly reduced when the Crow signed another treaty at Fort Laramie in 1868 under Chief Blackfoot, ceding 38,000,000 acres of which 8,000,000 would become the revised Crow Indian Reservation. This was later reduced further to 2,282,000 acres in 1905, which closely approximates the reservation's size in the present day (Frey 1987:30).

Many animals and birds are important to the Crow for subsistence, utilitarian, medicinal, sacred, and other uses. Wildschut (1975), for example, notes the particular importance of the Golden Eagle (*Aquila chrysaetos*) to the Crow for ritual and ceremonial purposes. Like many other Plains tribes, eagle trapping was a dangerous and very spiritual process practiced by the Crow to acquire feathers and parts. Trapping ceased with federal legislation that protects eagles from capture and harm. Golden Eagle feathers are utilized for ceremonial objects and regalia, including a holy pipe, and the eagle wing bone is used to make a whistle for the Sun Dance (Zedeño et al. 2006:120). Other significant birds identified during Zedeño and colleagues' ethnographic resource assessments include ducks, hawks, Great Horned Owls (*Bubo virginianus*), and swallows, all of which have medicinal and sacred significance.

CHAPTER THREE

What Makes a Bird?

Most would agree that the defining quality that sets birds apart from other types of animals is their ability to fly. Western biology defines birds using Linnaean classification, where the attributes of Class Aves (the birds) include warm-bloodedness, feathers, wings, a beak, the ability to lay eggs, and (usually) the ability to fly. From this broad avian classification, birds are separated further into orders such as Falconiformes (falcons, hawks, ospreys, and vultures), Anseriformes (waterfowl), Ciconiformes (herons, bitterns, ibises, and storks). Passeriformes (the largest order that includes flycatchers, songbirds, and doves), and so on. Each of these orders is then progressively broken down into families, genera, species, and subspecies. Folk taxonomies, on the other hand, include the same general concepts about what qualifies as a bird, but the classifications and methods of identification of birds vary across cultures. These taxonomies draw on everyday observations and experiences with the natural world and are fundamentally phenomenological (Rosaldo 1972).

Folk taxonomies hold varying levels of specificity concerning the identification of birds. Some may be referenced more generally, while others merit a very specific classification apart from any other bird. Language can provide a valuable perspective on the construction of folk bird taxonomies and reflect how people connect with and understand their surroundings (Zedeño 2008a). For example, Blackfoot language recognizes some small, neutrally colored songbirds as belonging to one group with a general name (*sistsi* and *póksistsi*, "birds" and "little birds"), while hawks might be broken down into six different species: Osprey (*Pandion haliaetus*; Plate 17), the Sharp-shinned Hawk (*Accipiter striatus*; Plate 21), Red-tailed Hawk (*Buteo jamaicensis*; Plate 22), Rough-legged Hawk (*Buteo lagopus*; Plate 23), marsh hawk or Northern Harrier (*Circus cyaneus*; Plate 24), and the American Kestrel or sparrowhawk (*Falco sparverius*; Plate 25) (Hungry-Wolf 2006:135, Schaeffer 1950:44). The Blackfoot also include bats (Chiroptera) in their avian classification, although according to Western taxonomy they are considered mammals, neither bearing feathers nor laying eggs, among other distinctions. This is but one argument against the idea that biological classificatory systems are independent from cultural transmission (Boster 1987). Although some birds are recognized more discretely than others, each kind of bird has a place and a purpose in Missouri River tribal culture, whether that role is small or large. The words used to classify birds are often loaded with cultural and social meanings beyond differentiating between two types of birds, and likewise the features and behaviors that mark a particular bird are framed in terms of local ontology and epistemology (Rosaldo 1972).

Local tribes distinguish birds through many qualities, such as color, body, behavior, and song. Native identifications of birds classify male and female birds of the same species in separate groups because of their different appearance. Conversely, they might conflate two similar looking birds of different sizes as one type of bird, even though they are two separate species in Western classifications (Hungry-Wolf 2006:135). Thus, local avian taxonomies rely on such phenomenological methods of identification, and this is reflected in the names given to birds and how they are described in stories or imitated in dance and ceremonial performances.

The Blackfeet Tribe has the most extensive linguistic collection of bird names available, thanks to two comprehensive sources: Walter McClintock's *The Old North Trail*

(1999), which documents knowledge of about 75 different kinds of birds collected over his time living among the Blackfeet at the turn of the twentieth century; and Claude Schaeffer's *Blackfeet Bird Nomenclature* (1950), which accounts for over 100 birds. While other tribes have less thoroughly documented taxonomies, classic and contemporary versions of oral traditions about birds reveal much about how native people conceptualize and classify birds.

COLORING

Some birds are distinguished only in a general way by their color. In the Hidatsa's explanation for the origin of rainbows, a bounded "red bird" is credited with causing a rainbow when it pounced on a hare but was held fast with fishing line tied to its claws, creating a semi-circle as it flew (Thwaites 1906:374). Other birds are clearly named for their particular coloring, often in a way that connects to some other facet of indigenous knowledge and experience. The Clark's Nutcracker (*Nucifraga columbiana*; Plate 33) is known as "old lodge-cover" or *Makokím* in the Blackfoot language because its color resembles the gray of an old and worn lodge-cover. The male American Goldfinch (*Carduelis tristis*; Plate 47) is named "grease bird" or *Pomisístsi* because its summer plumage resembles the appearance of the back fat on a bison (McClintock 1999:482; Schaeffer 1950:44). Others are known for their distinctive markings, like the Bobolink (*Dolichonyx oryzivorus*), which earned its Hidatsa name, the Skunk Bird, from the male's distinctive skunk-like black and white markings (Weitzner 1979:197). Similar to the English name, which recognizes its distinctive white coloring, the Blackfoot name for a Bald Eagle (*Haliaeetus leucocephalus*; Plate 19) is "white-headed" or *Ksikochkini*. The Gray Jay (*Perisoreus canadensis*), also known as the Canada jay or whiskeyjack, is known to the Blackfoot as *Apiakunski* or "white-forehead" for its pale gray coloring on the nape and forecrown.

Sometimes the meaning behind a bird's coloring or marking gives it important distinction. There are many stories in present traditions of Missouri River tribes to explain how certain birds came to look the way they do today, with contexts varying from sacred to humorous. In the Crow tradition, when Creator was making animals, he decided that the prairie-chicken[1] was too "dull" in its color and markings (Zedeño et al. 2006:201). He took red paint,

the most powerful paint color across Native America (Zedeño 2009), and lined the eyes of the prairie-chicken to brighten its appearance, also bestowing on it its characteristic dance at sunrise and sunset. According to one Crow consultant, this was the first use of red paint by Creator (Zedeño et al. 2006:245). Red paint also appears in one version of the Hidatsa creation story, where a hungry bird, which had just been created from mud stuck its head into red paint while looking for food, and became the Red-headed Woodpecker (*Melanerpes erythrocephalus*; Plate 39).

In another Crow story, the magpie[2] gained the white coloring on its wing as the result of a gambling hand-game instigated by Old-Man-Coyote between all the winged and wingless creatures (Lowie 1922a:236). Old-Man-Coyote alternately helped each group as they gambled for perpetual daylight or darkness. They played all night, and in the morning the birds finally won; magpie was the first to enter into the daylight, and as the sun struck his wing it became white and remained that way ever since.

The reddish-brown (rufous) coloring of the Sandhill Crane (*Grus canadensis*; Plate 11) is explained through the Assiniboine story of Crane and Otter. A mother crane asked Otter to watch over her son while she flew south, as he was yet too weak to fly. However, Osni' (Cold) came with winter, killing Otter and kidnapping the young crane to his home. The crane was forced to stoke Osni's fire with his bill, which caused the fire to burn his back and give his skin a rufous color, which persists today. The Sandhill Crane has a red crown and rufous coloring on its wings, and is especially pronounced in juveniles on the head, neck, and back (Sterry and Small 2009:112).

BODY

The physical features of the body are used in the identification and naming of birds, and often have a "just-so" explanatory account that describes how a certain feature came about. As mentioned previously, bats are included in regional bird taxonomies. Bats carry several names in the Blackfoot language in reference to their appearance and behavior. One example is *Motinstami* or "many-lodgepoles," in reference to bats' thin skin and protruding wing bones that resemble the lodgepoles of a tipi (Hungry-Wolf 2006:136; Schaeffer 1950:45). Pretty-Shield, a Crow

---

[1] Most likely the Greater Prairie-chicken (*Tympanunchus cupido*; Plate 16), whose range is farther north than its relative, the Lesser Prairie-chicken (*Tympanunchus pallidicinctus*) (Sterry and Small 2009:58).

[2] Refers to the Black-billed Magpie (*Pica hudsonia*; Plate 34), not the Yellow-billed Magpie (*Pica nuttalli*) that occurs primarily in California.

medicine woman interviewed by Frank Linderman, repeatedly referred to the Burrowing Owl (*Athene cunicularia*; Plate 27) as the "long-legged owl that lives with the prairie-dogs," referencing not only the bird's characteristic long legs in comparison to other owls, but also its habitat and behavior (Linderman 1972:43–44, 125). Two Blackfoot names for the American White Pelican (*Pelecanus erythrorhynchos*; Plate 8), *Soχkaúkomi* ("big throat") and *Atiisípissa* ("carries water") are also are descriptive of their physical characteristics and behavior.

The grebe (Podicipedidae; Plate 7) receives much attention in Blackfoot narratives and descriptions for its physical features, most notably its eyes and legs, which both have stories to explain their unique qualities. The Hidatsa explain the grebe's red eyes through the creation story where the bird dove so deeply for mud to create the earth that its eyes turned red (Wilson 1910 as cited in Wood 1986:100). The Blackfoot tradition holds that Old Man (Napi) had a hand in turning the grebe's eyes red (Grinnell 1913:185–187). Napi had tricked a group of waterbirds into dancing for him with their eyes closed as he sang, warning that if anyone opened their eyes they would turn red. He was killing each duck as it danced by him and throwing it into the middle of the dance circle to eat later. the littlest bird opened its eyes, saw what was happening, and warned the remaining bird. Grinnell identifies this bird as the Horned Grebe (*Podiceps auritus*), and the Blackfoot believe that ever since this incident the Horned Grebe has had red eyes.

There is a similar story to account for the strange posterior placement of the grebe's legs, which merit its Blackfoot name, *Apatsséssika*, which means "legs grow backward" (Schaeffer 1950:39). The narrative closely follows the previous account, but in this story when the grebe realized that Napi was wringing the birds' necks it tried to escape. Napi stepped on its legs, explaining their strange backward placement today (Hungry-Wolf 2006:138).

The Rolling Rock story of the Blackfoot explains why the Common Nighthawk (*Chordeiles minor*), referred to in the story as a bullbat, has such a small beak and a sizeable mouth. It is said that Napi was cold so he borrowed a robe from a large rock but would not return it. The rock got angry and began to chase Napi, so he called the nighthawks for help. After the birds destroyed the rock that was rolling in his pursuit, Napi rewarded the nighthawks by pulling their mouths wide and pinching off their beaks so that they were "pretty and queer looking" (Grinnell 1913:166–167). In another version of this story, Old Man asked the nighthawk for help, but subsequently changed his mind and accused the birds of spoiling his fun by destroying the rolling rock. He punished them by tearing off their bills and splitting their mouths open wide, accounting for their appearance today (Wissler and Duvall 1908:24–25). The nighthawks later retaliated by defecating on Old Man as they flew over his head. Today, Blackfoot elders ascribe great significance to archaeological sites where Common Nighthawks nest.

Supernatural birds are also identified by their distinguishing physical features, especially in relation to other common birds. Thunderbird is most often associated with the eagle, although the term sometimes represents all birds (Bowers 2004:260; Kennedy 1961:57; Lowie 1922b:322; Rodnick 1938:46). Hidatsa stories describe Thunderbird as having "a forked tail like the swallow, and its wingspan touched each side of the Missouri River," and contend that white-breasted swallows must not be eaten because they were believed to be Thunderbirds on account of their forked tails (Weitzner 1979:312). The Hidatsa supernatural being Fringe-wing, chief of the eagles, was also marked with unique physical features, such as tent-like skin covering her wings and body, fringes on her wings and tails, an eagle's beak and large size; however, she also takes the form of a "common eagle" (Wilson 1928:214–216).

BEHAVIOR

Local tribes consider behavior an important bird identifier, as certain birds often have unique behavioral qualities that set them apart from physically similar birds. Blackfoot use the word "*sae-eh*" (Hungry-Wolf 2006:135) or "*sée*" (Schaeffer 1950:40) as a general term for ducks and geese. The Common Merganser (*Mergus merganser*; Plate 2), however, is differentiated from the generic Blackfoot classification of ducks and geese by the addition of the prefix "*Mi-*" which means "hardy." The merganser is called *Misae-eh* because it remains in the northern Plains throughout the winter and because of its ability to remain underwater for extended periods of time (Hungry-Wolf 2006:135). The generic Blackfoot term for woodpeckers is *páχpaksksissi* or "pounding nose," obviously named for their drumming and drilling behavior on trees with their strong bills or "noses" (Schaeffer 1950:42). From this behavioral identification, woodpeckers are further sorted into more specific kinds by their coloring: the Red-headed Woodpecker is called *Mikimata* (fire-reddened-breast), or *Ekots-otokan* (red-head); the red-shafted Northern Flicker (*Colaptes auratus*; Plate 36) is called *Mik-anikesuyi* (flashes-red-feathers) for the brilliant red feathers under its black tail, and the yellow-shafted variety (Plate 37) is called *Otachkuyi-kanikesuyi*; the Pileated Woodpecker (*Dryocopus pileatus*; Plate 38) is called *Siksik-anikesuyi* (flashes-black-feathers) (Hungry-Wolf 2006:135).

The Wilson's Snipe (*Gallinago delicata*; Plate 14) has two differing accounts of its Blackfoot name and translation from Schaeffer and McClintock's reports, both relating to behavioral traits. McClintock recorded the snipe's Blackfoot name as "Shadow in the Water" (*So-otak-skan*), from its habit of frequenting shallow waters, where they can easily see their own shadow (McClintock 1999:483–484). Schaeffer's (1950:41) sources identified the Wilson's Snipe as "Rain maker" (*Sotamstá*) because of its "peculiar, circling, mating flight, which the Blackfoot say takes place during or after a rain," and which aligns more closely with field guide descriptions of its "rapid and zigzagging" flight pattern (Sterry and Small 2009:144). Such discrepancies are case in point for why correct field identification is vital to productive research, as it is hard to decipher the correct bird correlates between Blackfoot language and Western taxonomies from only brief textual excerpts.

Schaeffer suggests that perhaps McClintock's contradictory identification was rooted in the tendency for generalizations made among the Blackfoot when identifying shorebirds such as the snipe, plover (Charadriidae), and Killdeer (*Charadrius vociferous*; Plate 13), although the snipe's long bill would appear to be a noticeable distinguishing feature from other two short-billed shorebirds (Schaeffer 1950:41). Further issues arise when the consultant simply does not know the correct English translation of a bird that is familiar only in one's native language.

Prairie-chickens are also repeatedly branded in Missouri River traditions by their behavior, a conspicuous dance displaying fancy footwork as they rapidly stomp their feet into the ground and produce a booming call to attract a mate. The birds' dance is often imitated in dances and ceremonies, for example the prairie-chicken dance (*ciyō-wagadji'bi*) of the Assiniboine and Atsina tribes (Lowie 1909:56–57, 1960). The prairie-chicken also lends its name to several societies and clans that are known to imitate some of its other peculiar behaviors. The Hidatsa clan called the Prairie-chicken people (*Tsistska-doxpaka*), for instance, got their name from a war party that would withdraw from the rest of the group and reside in the buckbrush like their namesake (Wilson 1908 and 1911 as cited in Wood 1986:51, 106).

You ever see a prairie-chicken do its dance? Oh man, you're missing out. Some of the traditional dancers, the Grass Dancers, we try to mock those. The pheasant, we try to mock their moves. If they are in breeding season, you know, they puff their chest out or they shrug their neck, or they do their little dance with their feet. A prairie-chicken, his chest will pop out and this ultimate color will come out of him, and he'll just make a thumping noise [imitates the noise]. And there are so many. . . . When I first moved out to the country [in 1981], just over the hill where my cousin lives we used to call that Prairie-chicken Alley cause that's where they would come and dance during mating season. . . . I took my nephew and my brother with me and we went over the hill, and they were just out there, a whole bunch of prairie-chickens doing their dance. And we were looking at each thinking, you know this is something we'll probably never see again. . . . My uncle said, "Did you guys hear those? That's your dad's clan, the Prairie-Chicken clan, doing their dance."

—Mandan-Hidatsa consultant

The Blackfoot revere the American Bittern (*Botaurus lentiginosus*; Plate 9) for its behavior. Known variously as *Itapksissi natosi* (pointing beak to the sun) and *Natoipiksi* (sacred bird) (Hungry-Wolf 2006:135), the bittern is believed to follow the course of the sun with its beak as the day progresses. Schaeffer connects the name "sacred bird" to the bittern's behavior as a reflection of the emphasis on the Sun as an increasingly important spiritual element in Blackfoot culture (Schaeffer 1950:40). The bittern's behavior was mimicked in the Siksika's tobacco-planting rites in connection to the Beaver Bundle, which formerly held the skin of a bittern before it somehow disappeared (Hungry-Wolf 2006:138). When it was time to plant tobacco, the wife of the Beaver Bundle owner would dance in imitation of the bittern, directing her gaze toward the sun just like the bittern, and singing the phrase, "the sacred bird is powerful." The bittern's association with the sun and with water as a wetland species makes it an important part of the planting and growing processes, as it is believed to ensure good fortune and ample rain for the tobacco to flourish.

The Thunderbird, which is a central figure in Plains Indian cosmology, also received its name because of its behavior, manifested most often in thunder and lightning, as well as rainmaking. As both a physically and spiritually powerful being, the Thunderbird appears in many foundational stories for tribes in the region, including a story explaining the deeply rooted dichotomy between water monsters and Thunderbirds, in which a human helper received Thunderbird's power to shoot lightning from his eyes (to name only a few sources, see Beckwith 1937:92–95; Parks et al. 1978:55–57; Parks 1991:952–995). There are several explanations for how Thunderbird creates thunder, whether from flapping his wings (Denig 2000) or from the sound of his drummers (McClintock 1999:425–426; Parks 1991:400–405). The Thunderbird spirit makes his presence known to humans through this behavior, as well as through the physical manifestations of an eagle or other large raptor. Among the Blackfoot, Thunderbird is known to have stolen other men's wives and used lightning to blind the husbands, a behavior for which he was eventually punished by raven (Grinnell 1913).

BIRDSONG

And then four chickadees landed in a bush beside the walk, their busy little bodies giving me an idea. Every tribe of Indians in the Northwest respects the chickadee. They possess many stories about him and greatly enjoy telling them to friends. I would try to get Pretty-Shield going again by introducing the chickadee.

But I could not make our interpreter, Goes-together, understand which bird I meant. To her, as to most moderns, red or white, a bird is a bird. To these unfortunates there are "little" birds and "big" birds, and here their ornithology ends. I did not know the sign for chickadee, and during all the talking between Goes-together and myself Pretty-shield's face remained amusingly blank. At last, in desperation, I whistled the spring-call of the chickadee, and the day was saved. Pretty-shield reacted instantly. She stood up and with a hand resting on the table, leaved toward me, her eyes shining.

"Ahhh, Ahh! The chickadee is big medicine, Sign-talker," she said [Linderman 1972:151–152].

Some birds are named onomatopoeically for the sound of their call, like the small, gray, winter bird that the Hidatsa call *It-si-ki-ki*, in imitation of its sound (Weitzner 1979:198). Similarly, the Arikara word for goose[1] is ko-ut and ka'-ka is the name for the American Crow (*Corvus brachyrhynchos*; Plate 31), both approximating their respective calls. The Great Blue Heron (*Ardea herodias*) is *Moxkámi* in Blackfoot in imitation of its low, croaking call (Schaeffer 1950:39).

Other birds are recognized by their calls because they are similar to words or phrases in the local language, like a form of bird speech. The meadowlark was described in one Hidatsa story as "the yellow birds (who can almost talk Hidatsa)" (Wilson 1908:73–74). It is not a surprise that the meadowlark frequently appears in Hidatsa stories as a messenger. McClintock's Blackfoot consultant Brings-down-the-Sun shared numerous examples of this for a range of birds, including "a large woodpecker with red wings," *Ma-mi-as-ich-imi* (magpie), *O-toch-koki* (meadowlark), *Isik-o-ka-e* (Chestnut-collared Longspur), *Nepe-e* (White-throated Sparrow, *Zonotrichia albicollis*), *Pokaup* (Catbird), and the rooster, each of which has certain words or phrases of meaning attached to their song (McClintock 1999:481–484). For instance, the Gray Catbird (*Dumetella carolinensis*) is called "child" or "baby" by the Blackfoot because its call, na'á, so closely resembles the Blackfoot word for "mother" (Schaeffer 1950:43; McClintock 1999:482), and the Chestnut-collared Longspur (*Calcarius ornatus*) sings "Spread out your blanket and I will light upon it."

Some songs reveal a less agreeable temperament in the singer, however, such as when the meadowlark is known to occasionally wield unsettling remarks like Kitati-ma-siks-a-stoki: "Your sister has a black skin" (McClintock 1999:482).

The thing that's common about [the meadowlark], you'll notice how they sing in the morning. Sometimes they can almost say, it seems like they're almost saying, "Did you wash your face?" [laughs]. You know, how they sing and it sounds like they're talking?
—Blackfoot consultant

An Assiniboine story about the crow reveals that birds may also pay the consequences for their impudence in birdsong or "speech" (Blue Talk 1978). Originally, Crow was a brilliantly colored bird with an equally stunning singing voice. Unfortunately, he became arrogant and conceited, refusing to speak to Inktomi (a trickster figure in many Assiniboine stories) when he approached Crow in the form of an eagle. To punish him for his offensive behavior, Inktomi replaced Crow's beautiful birdsong with a strange cawing sound and took away his beautiful coloring, leaving him black.

On other occasions, birdsong is recognized not necessarily for its beauty but for its unique character and clarity. In two versions of a Mandan story (Beckwith 1937:74–75; Bowers 2004:280), birds lent their voices to others during a contest with an old woman for land. In the version of

---

[1] Likely the Canada goose (*Branta canadensis*; Plate 3).

the story recounted by Beckwith, a white bird from the goose family loaned its voice to an eagle, who won the battle against a buffalo bull for the loudest voice, while in the Bowers' version of this story a gull lent its voice to a man because it was known to be the clearest. These stories about birdsong begin to relate the personalities associated with certain birds, as do the stories and conclusions drawn from birds' behavior.

Birds are identified through a number of different avenues for folk taxonomies. Worldview, environment, context of interaction, and visual and aural cues all play a large role in how birds are viewed both literally and conceptually in the Missouri River Basin. Color, body, behavior, and birdsong are four fundamental means of identifying birds using local ecological knowledge. Each feature is integrated into oral traditions and indigenous belief systems creating a multi-scalar lens for the vitality and importance of avian life around the Missouri River.

CHAPTER FOUR

Birds and Origins of the World and the People

There was no land here before. It is said that there was nothing but water everywhere, so Iichíhkbaalee [First Maker] called four mallards. He told them, "I want to make land, dive into the water, and get me some mud. I will make dry land with that." One of the mallards went into the water but came up with nothing. The next one was the same, and so was the next one. Then the fourth one went into the water. He was gone for a long time and Iichíhkbaalee thought that the mallard had died, but then He saw the mallard coming with mud in his beak. He took the mud and made the earth as we know it today. He moved portions of water that today are the oceans, lake, and rivers. He molded the mountains, hills, valleys, and so forth. He made man also.

—Crow consultant (McCleary 1997:92)

HOW BIRDS CREATED THE EARTH

According to the oral traditions of virtually every tribe along the Missouri River, birds have been an integral part of life since the beginning of the world. Birds play a prominent role in stories about the creation of the earth, plants, and animals; the genesis of humankind and individual tribes; the creation of particular clans and societies; and the origins of certain ceremonies. In the Missouri River Basin, these storied traditions reflect the interaction among humans, birds, and other beings in a context where the boundaries between the earthly and the supernatural are blurred. Birds can take on an intermediary role between humans and otherworldly beings with their ability to occupy both the earth and the sky, thus linking the two worlds together (Murray 2009). In oral traditions, birds, humans, and supernatural beings are often able to speak to one another, undergo transformations, and take on supernatural abilities, defying the normal confines of everyday life. Birds, therefore, are helpers to both humans and supernatural beings. It is not a surprise that birds are often featured prominently in tribal origin stories.

The elemental role of the Missouri River and the waterfowl it attracts makes both of these things central to life in the region. The Missouri River is a powerful life force, providing wildlife habitats for fish, birds, and animals; vital nutrients for plants in the floodplains; and a source of holy water. Waterfowl such as ducks and geese figure prominently in riparian environs, and their aquatic adaptation may explain their persistence in oral traditions that include a flood event or a world covered entirely with water. The Mallard (*Anas platyrhynchos*; Plate 1), American Coot or mud hen (*Fulica americana*; Plate 10), hell-diver (grebe family), goose, and teal (Anatidae) are most often mentioned as helping create the earth by diving down to retrieve earth.

Not only are birds featured in almost every creation story, in many cases they were already around before the earth as we know it was formed. According to contemporary Assiniboine elders, the eagle was never "created"; it was always here, living alongside celestial beings. Waterfowl are often the only other creatures that existed at the time of creation. In the brief account of the Crow creation story at the beginning of this chapter, the Mallards help *Iichíhkbaalee*, the creator of the earth, in making land by retrieving earth from below the water. This is a recurring theme in Mandan (Beckwith 1937:1–2, 7; Peters 1995:24–25; Wilson 1908:93–95), Hidatsa (Bowers 1992:298; Parks et al.

1978:67; Wilson 1910 in Wood 1986), and Arikara (Dorsey 1904:11) creation stories, where birds similarly help the creator or creators (known variously as Old Man, Lone Man, First Creator, First Maker, and other names) in their work. In one version of the Crow origin story recorded by Lowie (1918:18), the ducks created the earth independent of Old-Man-Coyote, another form of the creator, before he became their advisor and molded the earth to create mountains, rivers, and all living things. The duck is also an important part of the Blackfoot's origin story involving a flood, where Napi sent ducks to the gather land from below the water. Blackfoot consultants note that duck bills, mallards specifically, are affixed to Blackfoot pipes to commemorate the important role of the duck in ancient times. These are central to the worldviews of Native people in the Missouri River Basin and explain the significance of birds in Native cosmology.

In the Hidatsa creation story told to Douglas Parks by John Brave, the world was jointly created by First Creator and Lone Man with the help of a duck (Parks et al. 1978:67). As is the case in most of the diving-waterbird origin stories, it takes several attempts—most often four tries—before the bird can successfully bring mud to the water's surface. The stress of repeatedly diving so deeply turns the duck's eyes red, explaining why the hell-diver (some kind of red-eyed grebe) has bright red eyes today (Wilson 1910, as cited in Wood 1986:100). First Creator and Lone Man use the mud to create the banks of the Missouri River and split the land for each to create living things on their respective sides. Another Hidatsa origin story recounted by Butterfly adds that after land was created on either side of the river, First Creator and Lone Man began by making grass, followed by trees and animals (Wilson 1910, as cited in Wood 1986). Lone Man and First Creator made birds on the fifth day.

When Lone Man created people, he made them from all kinds of plants and animals and therefore created a bond between people and particular beings. Native people often identify themselves as linked to birds since creation; some people consider themselves "eagle people" or "raven people" for example, or have an affinity toward other animals and plants such as bears and corn (Bowers 2004:365). John Brave's version of the creation story incorporates the Missouri River specifically, anchoring the centrality of the Missouri River region to Hidatsa creation cosmology. Several similar versions of the Arikara (Dorsey 1904; Peters 1995:26) and Mandan (Beckwith 1937; Wilson 1908:93–95) creation stories place the Missouri River at the epicenter of the world's creation.

BIRDS AND THE FLOOD

Many creation stories begin with a water-covered earth, but numerous North American Indians also have in common the story of another massive flood that covered the entire world (Krech 2009:169). The Mandan, Hidatsa, and Blackfoot have accounts of a second major flood event in their oral traditions, and in both cases the floods are linked to birds. Their ability to fly and their inclinations toward water facilitate their survival through these catastrophic floods and begin to explain their role in these stories, whether they are instigators or helpers during the deluge. In many cases, birds have the ability to cause a flood through their personal behavior, such as in the story to follow, in which Fat Bird's feathers held water that was released to cause a flood. In other stories, mammals or the trickster cause the flood, as in the contemporary Blackfoot version of the second creation, but birds can control the weather, or are allied with a force such as Thunder, which can cause it to rain and flood.

In one Mandan version of the flood, the people had been mistreating birds and animals in their hunting, for instance plucking all the feathers out of a bird they called Fat Bird and sticking its wing through the bill of the bird, greatly offending it (Beckwith 1937:18–21). The animals and birds planned a great flood to punish the people for their actions against the animals. A little boy overheard their plan and transformed himself into a magpie in order to spy on the animals and warn the people of the imminent flood. The Fat Bird's feathers were filled with water, which the animals used to cause the flood. In a slightly different version recorded by Bowers (1992), the placement of a feather through Fat Bird's bill causes the great flood. The magpie boy led his people to a place called Bird's Bill Hill, the highest hill in the area. He also saved his mother, Yellow Woman, who brought corn to the Mandan people. Only the magpie boy, his mother, and one man survived the flood and joined the people living on top of Bird's Bill Hill, which is known today as Eagle Nose Butte in Morton County, North Dakota.

In a similar Arikara story appearing in Wilson's field notes (1908:81–92), One Above, the creator of the earth and all people, flooded the world when his people decided that they were so strong they no longer needed him. The only living things to survive were a mosquito and a very small duck, which were able to subsist on the foam of the water and green water weeds respectively. When the duck brought the mosquito down into the water under his wing to feed on the water weeds, the flood receded and

One Above decided to give humans another chance. One Above recreated all the living things, but they remained underground until a mole broke through the soil and living things emerged on earth. This is a variant of the Arikara origin story elaborated below.

In an Assiniboine story, the recurring trickster character, *Sitcon'ski,* was duped by several animals and convinced Thunder that the animals were disrespecting them (Lowie 1909:110–111). As punishment, Thunder made it rain for a long time and flooded the earth. Sitcon'ski gave himself wings to fly up to the top of a high tree, and all animals except birds were killed in the flood. Sitcon'ski lived up on a mountaintop with Eagle, but after four days they became hungry and Eagle asked Thunder to stop the rain. After giving him an offering of dried buffalo meat, Thunder brought all the animals that had died in the flood back to life. This story demonstrates the communicative role that humans and birds play in traditional stories. Sitcon'ski is able to speak with Thunder, a supernatural being, as well as Eagle, and negotiates a massive flood. Similarly, Eagle and Thunder settle on how to end the flood.

Another flood story was recorded by George Catlin during his travels through Mandan country in 1832–1834, in which a bird, identified as a Mourning Dove (*Zenaida macroura*; Plate 35), brought a willow branch with fresh leaves to the Mandan people who had been travelling across the water for many days in a large canoe (Catlin 1967:30; Perrins 2009:288). Due to this sign of land, the people quickly arrived on the shore, where the bird proceeded to show them how to build their lodges, and to hunt and gather fruit. The Mourning Dove is regarded as a "medicine bird" (Catlin 1989:152), and its connection to the willow influenced the *Okipa*, an important buffalo-calling ceremony among the Mandan. In honor of the bird that showed them how to hunt, the Okipa ceremony did not begin until the willow was fully grown on the banks of the Missouri River (Catlin 1967:44). The willows themselves are considered sacred, as they hold the flood waters at bay. The Mandan pointed out the Mourning Dove to Catlin, as it could often be seen around their villages feeding on seeds that were embedded on the earthlodges' sod. They explained that the Mourning Dove was a sacred bird that should not be harmed.

EMERGENCE OF THE ARIKARA TRIBE

Birds also play recurring and important roles in the ethnogenesis of tribes and societies. The Arikara tribe's origin story illustrates the seminal role of birds in native genesis stories, while also incorporating the deluge narrative into their creation as a people (Dorsey 1904:12–17, 23–25). The world was originally occupied by giants, but *Nesaru*, the creator, wiped them out with a flood, saving the smaller people by placing them in a cave underground with Mother Corn, the mother of the tribe. Eventually these people wanted to search for a better world, so Mother Corn sent out four fast birds in search of a better place to live. Although they failed, Mother Corn and the Arikara people eventually emerged from the cave and began a journey westward, facing many obstacles as they went. At each obstacle they met, a different kind of bird was there to assist the people in overcoming the challenge. When they came to a large chasm, a Belted Kingfisher (*Ceryle alcyon*; Plate 30) used its bill to reshape the earth and create a bridge for the people.[1] They came to a thick and thorny forest, and an owl flapped its wings to clear the timber. Next, a Common Loon (*Gavia immer*; Plate 6) helped Mother Corn by parting a large lake with its speed, so that the people could cross and continue their journey west. At each place, the people had a choice to continue on or stay behind and turn into the bird that had helped them. When they had made it through their final obstacle, Mother Corn selected a prairie-chicken to carry a pipe to the Gods in all four directions, and to *Nesaru* above. She also explained to the Arikara that each direction is significant, and the Northwest quarter was the Wind and the breath of life, symbolizing all the powers of the air including birds and insects (Gilmore 1932:36–37, Will 1934:14). The spotted wings of the prairie-chicken are proof of how hard this journey was. The storm winds were so severe that as the prairie-chicken flew his wings became spotted white from the impact of all the rocks entrained by the wind (Dorsey 1904:20).

Several things become clear about the relationship between people, birds, and the supernatural from this version of the Arikara origin story. Birds often take on the role of messengers and helpers in these stories; Mother Corn sends out her fastest birds to scout out new territory

---

[1] In an alternate account of the Arikara origin story acquired by Wilson (1908:85), the bird that helps the Arikara cross the chasm is "a bird with a sharp bill, the most sacred of all birds." Its behavior is further elaborated in the statement that "whenever it strikes its beak it makes a hole" (Wilson 1908:85). From this description it seems possible that this unidentified bird is a species of woodpecker, although woodpeckers are not referenced elsewhere as the "most sacred bird." The majority of sources mention a kingfisher in this story, and the presence of the kingfisher in one of the Sacred Bundles of the Arikara further suggests that the kingfisher is more likely the bird intended in the origin story (Gilmore 1932:45).

and report back to her. Their journey surely could not have been completed without the help of the kingfisher and the owl in the Arikara migration to their new lands in the west along the Missouri River (Dorsey 1904:35). Furthermore, birds do more than just help people: They are also helpers to supernatural beings like Mother Corn. Each bird is bestowed with the skills and power to assist Mother Corn in helping the Arikara to overcome seemingly impassable obstacles, and the prairie-chicken is chosen by Mother Corn to offer pipes to Gods in each of the four directions. Thus begins the tradition of making an offering of smoke to the gods through a pipe, one of the most important and pervasive traditions of Missouri River tribes.

THE HIDATSA EARTH-NAMING CEREMONY

> Looking up [Raven Necklace] saw that Owl was speaking. Owl said, "This valley is known as Owl Valley. You can make a buffalo corral here. I will give you a ceremony called Earthnaming. When you perform the rites the other spirits will teach you the songs and what things to use with the Earth medicines. When you call them together, they will tell you the names of these high hills. There will be a great deal of memorizing [Bowers 1992:434].

Hidatsa territory was essentially defined by the boundaries of important buttes, the names of which were revealed to Raven Necklace by various spirits who each lived in a butte (Bowers 1992:434). Raven Necklace was an Assiniboine boy who was taken prisoner by the Hidatsa. The leader of the spirits was the owl,[1] who first appeared to Raven Necklace as he was about to push over the tree that the owl lived in. Owl explained that it was important to know the names of the buttes and the spirits who lived in each one, because they would help him learn the songs and uses of certain medicines. Owl lived in the Killdeer Mountains, and because he was the leader, this is where the spirit animals of the buttes gathered when they all came together. Other spirits lived in various buttes, which became important to the Hidatsa cultural landscape, including "Ghost Singing Butte situated northwest of the Killdeer Mountains, so named because Swallow and Hawk were buried there; Crow Butte; Singing Butte;

Heart Singing Butte; Little Heart Singing Butte; Fox Singing Butte; Rosebud Butte; White Butte; Opposite Butte; Buffalo Home Buttes; and others not now remembered by Hidatsa informants" (Bowers 1992:435).[2]

Although the particular histories and information about all of the buttes given by the owl and other spirit birds and animals are no longer remembered, the buttes are a culturally significant part of the landscape for the contemporary Hidatsa. It is important to note that of all the buttes listed by Bowers, Ghost Singing Butte is one of the few buttes mentioned by consultants for which a significant history and understanding still remains. Incidentally, this butte is also associated with two birds: the swallow and the hawk. They are believed to be buried in the butte and their spirits reside there. The swallow and hawk are known for bringing horses to the Hidatsa by appearing to a poor Hidatsa man and his wife; the swallow and hawk informed him that they would "bring horses down from the heavy timber to replace the dogs" (Bowers 1992:435). The birds explained to the man and woman how to use the horses and care for them, as well as the instructions for the Bird ceremony to honor the swallow and hawk who resided in one of the earth-naming buttes.

ORIGINS OF BIRD CLANS, SOCIETIES, AND CHAPTERS

> In the founding myth, a Goose spoke to the Arikara as follows: "I will go to the edge of the big rivers. When it is time for you to prepare something for me to eat, I shall return. When I shall have come back, you may proceed with your garden work, and you will be sure of success." That is why the geese always came in the spring, when the sowing began, and why they departed after the harvest [Mails 1973:150].

Many other societies, ranging from informal to very sacred, have their origins connecting back to birds. The Blackfoot Okan or Medicine Lodge ceremony is one of the most revered ceremonies, given to their culture hero Scarface by the Sun after he killed vicious birds who were attacking people in Sun's land (Grinnell 1913, 1920:95–103; Wissler and Duvall 1908:61–65). Scarface received two raven feathers to use for the Medicine Lodge ceremony, because the raven is the "smartest" of all (Grinnell

---

[1] The owl from the Earth-naming ceremony and rites is a "speckled owl," according to Bowers. It is unclear what species this refers to. The claws of the "speckled owl" are part of one of the Hidatsa Earth-naming bundles (Bowers 1992:438).

[2] See Map 1 of Bowers (1992:12) for the locations of these buttes and the tentative boundaries of Hidatsa territory.

1920:101). The Crow-Painted Lodge of the Blackfoot came about when a man was eagle-trapping on a hill and dreamed of a crow, who then gave him the lodge and taught him the songs and ceremony to go with it (Wissler and Duvall 1908:95).

In a ceremony Will (1934) refers to as the Arikara "medicine lodge" each clan performs as they enter the sacred lodge, including several bird clans (Will 1934: 17–18). Will (1934) mentions the Duck Clan and Owl Clan, along with a clan known variously as Crane or the Bald Eagle clan, which he suggests actually represents a cormorant based on its description. Other more informal societies also bear the names of birds, such as a dancing society in the Hidatsa tribe created for young boys called Magpies (Bowers 1992:180).

ORIGINS OF NATURAL PHENOMENA

Birds are credited with creating many natural features and phenomena as the world became the way it is today, such as seasonality, night and day, and weather. For example, in the Assiniboine traditions of Inktomi, birds are strongly linked to two stories about the seasons and how they came to be. A long time ago, supernatural beings living in the sky asked Inktomi to help them steal summer from a man who kept it in his lodge beyond the snowfields in the east (Lowie 1909:101–105). In this story, birds act as guards of the man's lodge and as helpers to Inktomi as he attempts to steal summer. Kata'pknaa'dan, an owl-like bird with the ability to fly noiselessly, was sent by Inktomi to scout out where the summer was kept in the lodge. Then, with the help of Wolf, Coyote, Red Fox, Jack-Rabbit, and finally Teedan' (a fast-flying, hawk-like bird), they evaded the owner and his guard birds and successfully brought summer to the beings in the sky. This ended eternal winter, and Inktomi and Frog decided to split the year between winter and summer. The Arikara have a similar story for the origins of the summer season, where Raven, Coyote, and Scalped Man stole the Sun's child to lure the Sun northward. Raven and Coyote carried the Sun's child as far as they could, until they had to rest and some birds came for the Sun to retrieve his child. This spot formed the boundary for winter and is also where the Arikara live today (Parks 1996:123–125).

The crane is known to have brought about the existence of four seasons later on, according to Assiniboine traditions (MWP 1942:30). Whereas there had previously been only two seasons each year after Inktomi and his animal helpers stole Summer, the headmen of the Assiniboine decided that cranes should carry Summer back and forth as they migrated over the year. When men took Summer south with them, they travelled so quickly that Winter came on in stark contrast to Summer. However, cranes were the first of the migratory birds to go south, and they moved gradually, with many stops along the way to feed. Winter would follow them as they moved, and that transitional time would be called "Fall"—*Pdanyedu*. As the cranes returned northward, plants and animals would show signs that Summer was on its way back; this time was called "Spring"—*Wedu*. In another version of this Assiniboine story, Frog travels east to find Spring and then wraps it into a bundle and sends it on the backs of two cranes to bring spring to the west (Denig 2000:220).

Birds are also associated with introducing other natural phenomena. The Hidatsa associate a "red bird" with the origin of rainbows, as the bird tried to fly with his feet tied together to a fish-line, creating an arc as he went (Thwaites 1906:374). The Bald Eagle, who is also the Thunderbird in the cosmology of many Missouri River tribes, causes powerful thunderstorms in Spring and early Summer because his enemy, the bear, arrives as the constellation Bear Above in Spring and disturbs Thunderbird (McCleary 1997:23–24). Similarly, summer clouds are thought by the Hidatsa to be produced by big birds as they searched for snakes and water spirits (Bowers 1992:375). Lightning is seen as a contest between these opposing creatures as they clash. Swallows are called fire birds by the Blackfoot because they are believed to have brought the first fire to Indians (McClintock 1999:483).

There are stories about the creation or naming of certain landforms throughout the history of the tribes in the area that reference birds, especially hilltops and buttes (see the previously mentioned Hidatsa Earth-naming Ceremony). For example, Table Butte (*Duhu ihgishi Sh* or "Thunderbird's Nest" in Hidatsa) is known as the place where the Thunderbird had its nest, as well as the origin place for the Hidatsa Low Cap Clan (Garcia 1995). It is there where Packs Antelope saved juvenile Thunderbirds from being eaten by a snake. In an Arikara story, a woman brings her baby with her to the prairie to pick turnips (Parks 1991:615–617). She lays the baby down as she moves further into the prairie to gather roots. When she returns, her baby has transformed into a hawk flying in the sky.[1] Trying to coax her baby back to the ground, she exposes

---

[1] In Wolf Chief's Hidatsa version (Wilson 1909:26) of this story, the bird was a Long-billed Curlew (*Numenius americanus*), which is supposed to make a sound like a baby crying.

her breasts to see if the child is hungry. But the hawk, her baby, continues to fly higher, and her breasts become two pointed hills called Young Woman's Breasts Buttes by the Arikara. The area today is called Virgin's Breasts Buttes, an adaptation of the original Arikara name.

BELIEFS ABOUT BIRDS AND BABIES

Buffalo Bird Woman, a Hidatsa informant to Wilson (1911:207), explained a belief that the spirits of birds and other animals were sometimes born to humans. They came from a different place than other babies, and would manifest themselves from within a person's body when they got older.

> We believed that birds and animals came and entered into women in some way and were born as their babes. Two Shields and my husband Son-of-a-Star, after they grew up remembered the very camp whither they had come as spirits to choose mothers and had there entered them to be born. These birds and animals that chose mothers did not come out of the Children's House, or Babies' House. They came from a different place, or places. . . . All babies born among us did not come from birds and animals and other things [Wilson 1911:207–212].

This belief shows a connection between birds and humans that even suggests a bond of kinship. In effect, a mother gives birth to the human embodiment of a bird spirit, who has selected her as its mother. A woman might not be aware of her unique circumstances until later in the child's life when they begin to show signs that they hold a bird spirit. Buffalo Bird Woman continued that if a child died before its teeth had grown, it would become or bird or animal again or go back to a place like the Babies' House (a landform also known as Babies Butte), where all the baby souls reside and where women go to ask for a child.

In sum, birds are omnipresent when things are created, whether they retrieve the first bits of earth; help create people, tribes, societies, clans, and important landforms; or are responsible for the origins of natural processes like the seasons and births.

CHAPTER FIVE

Bird Qualities

When you see different birds flying around, chasing each other in the air, it's because they are communicating with each other, they are engaged in some spiritual battle, some spiritual doing. It is very significant.

— Blackfoot consultant

People often attribute human qualities to the natural world (e.g., plants, animals, birds, etc.) as a way of relating to our surroundings. Humans bestow personalities, motivations, and emotions upon the actions of living things or objects even when scientific explanation offers alternate accounts of these experiences. These personifications vary across cultures and regions, and are often rooted in oral and written traditions that reach far into the past and draw upon a blend of the natural and supernatural worlds. Although in Western worldviews these ascriptions are merely anthropomorphisms or personifications, in Native North American ontology they reflect the recognition of plant, bird, or feather, for example, as an animate object or "person" in its own right (Hallowell 1976; Ingold 2000; Zedeño 2008b; 2009). Plants, animals, objects, spirits, and gods may possess certain personality traits, emotions, and abilities outside the common realms of nature. A nonhuman can take on many forms, shifting from a spirit to an animal, a rock to a bird, and man to a plant, or any combination of metamorphosis, and may take on new personalities or skills as they change form.

PERSONALITIES AND EMOTIONS

Just like people, birds communicate their personalities and emotions through a combination of body language, voice, and behavior. Local communities become acquainted with the personalities of different birds through observations of repeated actions, a "phrase" of birdsong that carries meaning, or a specific experience or interaction with a bird. Bird personalities may further carry an intrinsic cultural meaning that is not readily obvious to outside observers, founded in the storied interactions between birds and people, culture heroes, and supernatural beings in oral traditions of the distant past as well as the present. The kingbird[1] is known to Blackfoot women as "Stingy-with-his-berries" because the bird becomes very angry and agitated when women come around to gather berries and disturb him (McClintock 1999:483). Kingbirds are noted in field guides for their aggressive behavior, defending their nests with rigor and attacking other, larger birds like hawks and crows. Along these lines, Schaeffer's sources suggested that the kingbird's Blackfoot name has its origins in a story in which the kingbird successfully expelled a bear from its berry patch (Schaeffer 1950:43).

The nighthawks are a common feature in tribal stories that highlight their personality, nesting behavior, and odd appearance. For example,

The nighthawk is thought to be lazy because instead of building its nest in a tree like most birds, it nests on the bare ground. This personality trait also appears in a Mandan and Hidatsa Coyote story when the trickster Coyote gets his nose stuck in a corn-ball grinder and implores the help of some nighthawks flying above.

---

[1] Three kingbirds—Eastern (*Tyrannus tyrannus*), Western (*T. verticalis*), and, less commonly, Cassin's (*T. vociferans*)—are present in the Missouri River Basin.

Insisting that the corn-grinder had insulted the birds by calling them lazy and strange looking because of their nesting habits and small beaks, Coyote convinces the nighthawks to help him free his nose by destroying the grinder [Beckwith 1937:293].

The Killdeer is also a ground-nesting bird, which leaves its young particularly vulnerable to predators. Killdeers are known to go to great lengths to distract potential predators away from their nests, feigning injury or using a threat display to distract or frighten intruders (Brunton 1986). Blackfoot observations suggest that the Killdeer expresses immense sadness if its nest is robbed of its young, and "it grieves so deeply, and cries so hard, it will fall upon the ground" (McClintock 1999:483). This characterization draws heavily on both the behavior and birdsong of the killdeer, relating these actions to the human emotions associated with the loss of a child.

The meadowlark's personality is directly expressed through birdsong according to the Blackfoot interpretations of its call. The translations of the meadowlark's bird calls consist of several phrases: *kinakíni otsitotsínnaiyixpi*, "where there's fat on the liver"; *kitákema (nitákema) siksistókiwa*, "your (my) sister's black skin"; *ekyotsístsini*, "roof of mouth is red"; *matsiíkiwa sikimákewa*, and "good whistler (is a) stingy woman" (Schaeffer 1950:43). Each phrase is meant to be slightly offensive, showcasing the bird's cheeky personality. Despite its impertinence, as one of the earliest arriving birds in the spring, the meadowlark is lauded and greeted as an old friend.

The magpie is equally noted for its impish personality by the Crow; they are known to steal away meat left to cure on meat-racks (Linderman 2002:13). Young boys are associated with magpies because they tend to share their mischievous and persistent personalities. There is even an informal dance society for young Hidatsa boys called the Magpies (Bowers 1992:180).

Several birds are known for their wisdom and sharpness of mind, including the chickadee, crane, eagle, and raven. The chickadee was the medicine bird of Plenty Coups, a noted warrior and the last hereditary Crow chief; its powerful qualities are known to many Crow through its relationship with Plenty Coups (Linderman 2002; Zedeño et al. 2006:224). In a dream, Plenty Coups learned about the character of the chickadee as small but wise, and perhaps almost diplomatic:

"Listen, Plenty-coups," said the voice, "In that tree is the lodge of the Chickadee. He is least in strength but strongest of mind among his kind. He is willing to work for wisdom. The Chickadee-person is a good listener. Nothing escapes his ears, which he has sharpened by constant use. Whenever others are talking together of their successes or failures, there you will find the Chickadee-person listening to their words. But in all his listening he tends to his own business. He never intrudes, never speaks in strange company, and yet never misses a chance to learn from others. He gains success and avoids failures by learning how others succeeded or failed, and without great trouble to himself. There is scarcely a lodge he does not visit, hardly a person he does not know, and yet everybody likes him, because he minds his own business, or pretends to.

The lodges of countless Bird-people were in that forest when the Four Winds charged it. Only one is left unharmed, the lodge of the Chickadee-person. Develop your body, but do not neglect your mind, Plenty-coups. It is the mind that leads a man to power, not strength of body" [Linderman 2002:37].

The chickadee is judicious, and a strong ally in times of war as well as peace because of its counsel and wisdom gained from the experiences of others. Chickadees can also be very clever, as in the Hidatsa story of the Two Twins, when the boys turn themselves into chickadees to trick their enemy Shoulder Mouth into swallowing a scalding rock to defeat him (Wilson 1908:23–24, 64). A Blackfoot elder recounts,

My grandfather's name, Joe Birdrattle derives from a chickadee rattle transferred to him by a Kootenai man known as Eneas, and that he kept in a bundle. As part of the bundle [rules] Joe Birdrattle was restricted from attending pow-wows and one day he broke that restriction. When he opened the medicine bundle afterwards the rattle flew away in the direction of Kootenai land.

The eagle is also revered as "sacred, very wise, and almost-human-like" (Hungry-Wolf 2006:137). The crane is known to the Crow as the cleverest of animals (Lowie 1918:297, 1919:164). The Common Raven (*Corvus corax*; Plate 32) is also noted for its cunning personality, as well as its smarts. The raven is an opportunistic feeder and will also hunt young animals in pairs, with one bird distracting the adult while the other steals away eggs or newborn birds and animals (Cornell Lab of Ornithology

[CLO] 2011). In a Blackfoot story, the raven's resourcefulness and clever nature are recognized by the Sun, who points out, "What one of all animals is smartest? The raven is, for he always finds food. He is never hungry" (Grinnell 1920:101).

POWER AND SKILLS

Some birds take on specialized roles in Missouri River society and cosmology for the specific abilities and skills they possess. These abilities may be utilized by birds for their own purposes, as well as in their interactions with humans and supernatural beings. Special bird attributes and capabilities usually become apparent through oral traditions, but are strengthened by personal experience, dreams, important bird-associated material culture, and ceremonies—all of which confirm the power or medicine that resides within a certain bird. Some birds are considered augural; they might signal good or bad luck, a certain type of weather, the presence of enemies nearby, or some other omen. Others are known for their particular expertise: strength, bravery, speed, success in war, hunting abilities, skill in farming, and so on.

One of the most significant distinctions between native and Western taxonomies is metamorphosis, or the ability of a particular thing or being to become another. Oral traditions and personal accounts illustrate many forms of bodily transformations. This metamorphosis can happen in many directions: a person might be transformed into an animal, bird, spirit, or even an object like a rock or an arrow. Alternatively, an animal or a rock might take human form, a spirit might take object form, or any combination thereof. It is this fluidity that lends animating power to the natural world.

The Hidatsa story of Burnt Arrow, which in many ways parallels the Arikara tradition of Speckled Arrow and the Mandan tradition of Charred Body, exhibits several transformations by the central character and how these changes affect his powers and abilities. Burnt Arrow is first described as a man who lives in the sky, but he comes down to earth in the form of an arrow (Wilson 1908). The arrow was marked with three lines, which denoted that Burnt Arrow was also a bird, specifically thought to be *Ic-pa-ta-ki*, the Hidatsa word for a bird with a white tail and the body of a hawk (Wilson 1908:1). He is often referred to as a Thunderbird (Wilson 1911). By taking his bird form he is able to return to the sky world through flight. In the story of Burnt Arrow, he wants to return to his earthly village quickly so he takes the form of an arrow;

however, he becomes stuck in the ground and cannot free himself.[1] Even in his arrow form he retains the powers associated with birds, such as rainmaking, and through this he is able to drench the soil until it is soft enough to free himself (Wilson 1908:3). In some versions of this story, the arrow becomes 20 lodges of Indian people, corresponding to the 20 parts of the arrow, and this is one way that the origin of Native people is explained (Wilson 1908:72).

Physical Attributes

The hawk is noted for its speed or "fleetness" (Lowie 1918:183–184). One story that persists throughout oral traditions of the tribes in the region involves a man who comes across a rabbit and a hawk in the woods (Lowie 1918:183–184; Parks 1991:967–969; Wissler and Duvall 1908:122). Pursued by the hawk, the rabbit promises to give him some kind of power (speed, long life, success in hunting), while the hawk offers his speed to the man. Depending on the storyteller, the man refuses one or the other animal, or makes a compromise by catching a squirrel for the hawk to eat instead, thus gaining the power of the rabbit, the hawk, or both.[2]

The Hidatsa revere the eagle and the hawk for their speed and endurance, which allows them to overtake their prey (Wilson 1911:154). The men of the Hidatsa's Stone Hammer Society attached eagle and hawk feathers to their ceremonial stones, and prayed to the feathers of these birds to imbue their speed and strength on the Stone Hammer Society members. One elder member of this society explained to the newly initiated members:

> Our sons, now the society is yours. Everything that belongs to the society, belongs to you. But these feathers that we have tied on the stones, we believe that we should pray to them, we Stone Hammers, when we go to war. As hawks and eagles, whose feathers these are, go faster than any other birds, so also may we. As these birds capture other birds that they pursue, so may we. You therefore must use this same custom, and pray to these feathers, to make you strong in war against your enemies, and to make you strong to capture your food! [Wilson 1911:154].

---

1 Burnt Arrow's village is referenced by Goodbird as near a coulee somewhere around Washburn, North Dakota (Wilson 1908:47).

2 The "hawk" in this story is actually a falcon, the American Kestrel (*Falco sparverius*). It is sometimes known as a sparrowhawk.

Magpies are sometimes named as the fastest of all birds, even over the eagle and the hawk. In a Mandan story, an eagle-man named Looks Down from Above was challenged by his mother-in-law to a race, the prize being all the buffalo in the world (Beckwith 1937:66–76; Lowie 1918:114–115). The eagle chose to transform himself into a magpie to run the race, noting that it was the fastest because it appeared to draw its target toward itself with every swoop of its flight. With the magpie's fleetness and help from a curlew[1] and a duck, he won the race and was entitled to all the buffalo. However, the old woman, wanting to keep some of her buffalo, told Looks Down from Above that he could keep the buffalo as far as he could see. Looks Down from Above enlisted the help of a raven, who is known for its excellent vision. The raven sometimes would lend its "long vision" to men by allowing them to use his eyeballs to see across long distances (Bowers 2004:280). With the help of the raven's eyesight, he could see the entire earth from the top of a high hill, and ordered his mother-in-law to release all the buffalo she had kept trapped inside the earth with a boulder. The buffalo were now free to roam the land as food for his people.

Ravens are similarly recognized by the Blackfoot for their excellent vision. In Grinnell's (1913:53–59) account of the origin story for the Thunder Pipe (Medicine Pipe), a man's wife was stolen by Thunder. A raven invited the man into his lodge and offered to help him defeat Thunder and get his wife back. Raven is considered to be the only one who is more powerful than Thunder. When the man doubted Raven's power, Raven took him outside and, rubbing some medicine on the man's eyes, enabled him to see his camp that was located many days' travel away from Raven's lodge. With renewed confidence and help from Raven, the man was able to defeat Thunder and in the process he received the Medicine Pipe of the Blackfoot.

Seasonality and Weather

Every month the Beaver Bundle people would watch for the different types of birds that start coming back. So during the eagle month, the eagles were coming back, the Beaver Bundle guy would sing the eagle song for that month to signify the time of the year. Every month there would be different birds coming back, and they all need to hear those songs. They are so important to them. Just like nowadays, how we all kind of basically know a little bit about the computer. For our

people back then it was so necessary that we knew nature inside out because that's how we survived, you know? We knew how to tell the seasons and tell the weather from watching the environment, and the birds. It so significant and important for our group to know the environment.

—Blackfoot elder

Thunderbirds and Eagles

For the Mandan and Hidatsa, "Big Birds" as a category refers to eagles, hawks, ravens and crows, all of which are considered connected to the Thunderbird spirit (Bowers 1992:363). This classification of birds seems to extend to all the tribes of the Missouri River Basin, although at times it may be confined to raptors, or extended to consider all birds as representative of the Thunderbird (Rodnick 1938:46). Big birds, and the eagle in particular, are credited with the ability to cause rain and thunderstorms, as the Thunderbird is the god of rain (MWP 1942:221). Thunderbirds may be called upon during the Big Bird rites of the Mandan and Hidatsa in association with both rain and warfare (Bowers 2004:108). The approach of a rainstorm or the sound of thunder signals the presence of the Thunder spirit. If this occurs during a ceremony or dance related to the Thunderbird, it may be seen as Thunder arriving to collect his offering (MWP 1942:223).

The Thunderbird is known to share his power to control the weather with some individuals who are visited by the Thunderbird spirit in a vision or dream, or with important cultural heroes in traditional narratives. There is a story well-known among the Mandan, Hidatsa, and Arikara tribes in which a man (known variously as Packs Antelope, Carries the Antelope, and Tied Face) is carried up into the nest of a Thunderbird, who asks him to kill a water monster in order to save his children (Beckwith 1937:92–95; Dorsey 1904:78; Parks 1996:206–209). In Good Bird's relation of the story, as a reward for his help Packs Antelope was turned into a Thunderbird (Wilson 1908:133). The Thunderbirds instructed him to lie in the nest, and Packs Antelope was turned into an egg. Four Thunderbirds brooded on the nest, and after the last one finished Packs Antelope was a full grown Thunderbird. Later, when Packs Antelope was turned back into a man, Grandfather Snake forgot to heal his eyes, so that when he looked around he shot lightning from them (Wilson 1908:136). In other versions of the story, the man's ability to flash lightning from his eyes was a power bestowed to him by the Thunderbirds, along with the ability to control rainfall (Beckwith 1937:92–95; Dorsey 1904:78; Parks

---

[1] In Lowie's (1918:115) account this is a snipe.

1996:206–209). When he returned to his village Packs Antelope gave his people several instructions before he would go back to live with the Thunderbirds. Packs Antelope advised on behalf of the Thunderbirds:

> If I come in the storm a long way off from you and you are out on the prairie or on a hill and you see me coming and you give me [a] sign that I must go in another direction I must do it. And if rain or hail came and one of my band [the Lowered Caps] motions with his blanket like the wing of a bird and so motion for the storm to go around it will do it. This is because I belong to the Thunder Birds. If a crash of thunder comes very loud, you will know that it is a sign that enemies are near. Beware then! Look out for yourselves! It is my warning! [Wilson 1908:137].

In another remarkable encounter, a Crow man named Lone-tree became separated from his group while fleeing from a war party. As a storm approached he was visited by "a big bird coming down from among the clouds" (Lowie 1922b:330). The eagle alighted on the ground across from him, protected him through the hailstorm, and promised to adopt him, telling Lone-tree he was the Thunder spirit. Lone-tree made a necklace of large white stones to symbolize the hailstones, and henceforth was able to harness the power of lightning, thunderstorms, and hail.

In addition to controlling rain and storms, eagles also have the ability to control the wind; this power was used in hunting to switch the direction of the wind so that game would not smell a hunter approaching (Lowie 1918:170). Often the power to control the wind was contained in an eagle's feather, which was a gift from the eagle to its ward. Two Leggings, a Crow warrior, possessed one such feather, which was painted with six white spots (Nabokov 1970:159). The wind would blow from whatever direction he pointed the feather at, and the painted white spots on the feather gave him the ability to cause a hailstorm between himself and any enemy who was chasing him.

Ravens and Crows

The Blackfoot regard the Raven as even more powerful than the Thunder spirit (Grinnell 1920:113–116; Wissler and Duvall 1908:134). Thunderbird and Raven once tested their powers against each other when Thunderbird stole Raven's wife and refused to release her back to Raven (Wissler and Duvall 1908:134). Raven used his medicine to bring a strong winter storm upon Thunderbird. Thunderbird was forced to flash his lightning constantly in order to keep warm

Figure 5.1. Blackfoot Snow Tipi. Source: McClintock (1999:34).

and melt the snow around his body. Eventually he gave up Raven's wife. For this reason, the Raven has the ability to bring on cold weather and snow; people will plead with the Raven during particularly bad winter weather to take mercy on them and stop the storm (Figure 5.1).

> When we were little, we were told the origin story of when Thunderbird and Raven fought for their home, for Thunder's home. One of those mountains by Crowsnest Pass, it was originally called the Crow's home. Anyway, Raven and Crow fought over that point and Thunderbird opened his eyes and would shoot lightning bolts at Raven. Raven would flap his wings, and that froze his eyes shut so he couldn't use his lightning anymore. So Thunderbird gave him his home. . . . After that, when we were kids, when there was a small thunderstorm coming toward us, as a kid you would run out and start cawing like crows. And all of a sudden that thunderstorm just moves around you, because he remembers that story.
>
> —Blackfoot consultant

The Blackfoot consultant continued that if one sees a raven exhibiting a particular type of behavior, this is a signal that weather will change: "When you see a raven in the winter time and he's shaking his wings, it means he's making cold weather. So it's kind of like, we get ready for cold weather." Wissler (1912:82) was told by a Blackfoot medicine man that, "[o]nce in a dream I saw some crows with white paint on their breasts and tails. The crows told me they would give me power, so that when I wished the weather to be foggy I must paint myself similarly. Thus, I got power to make foggy weather. Also, they gave me power to get much food, property, etc."

In the Crow story of "The Dwarf's Ward," a young man was given raven feathers by his parents as protection (Lowie 1918:165–169). He used these feathers to send out a call for help when his new wife's seven brothers tried to kill him with lightning. The raven feather was sent up through the smoke-hole of the tipi, and was answered by a raven whose "caw" brought a severe blizzard to the tipi. His enemies were frozen stiff, and only the young man and his wife were able to resist the cold by huddling together. Again, he sent out the feather to tell all of his helpers that the brothers were frozen; the feather came back as a cawing raven.

Cranes

In Assiniboine tradition, cranes are tasked with carrying summer back and forth with their migrations north and south (MWP 1942:30). As one of the earliest birds to begin migrating south in the fall, they carry the warm weather with them. The seasons change gradually as the cranes progress, stopping periodically to feed at different spots along the Missouri River and other wetlands as they travel. This accounts for the existence of all four seasons. The variance in the duration of each season is explained by the quality of the feeding grounds and the speed of the cranes traveling south or north, with a late spring signifying that perhaps the cranes lingered at a particularly rich feeding area for longer than usual.

Geese

Geese are recognized as being especially attuned to the weather and are able to predict changes (Grinnell 1920:260). The Blackfoot say that when the geese are flying high it is a signal that winter will come early (McClintock 1999:126–7).

"Red-Shouldered Oriole"

During times of drought, the Hidatsa used special pots given to them by Old-Woman-Who-Never-Dies for rain-making rites. Maximilian referenced the "red-shouldered oriole" as a medicine bird associated with those rainmaking rites because it was seen emerging from the water, dancing and singing as Old-Woman told the people about the pots and associated rainmaking procedures to be followed (Bowers 1992:348). The "red-shouldered oriole" is an old name for the Red-winged Blackbird (*Agelaius phoeniceus*), a common bird in watered areas in the region. The characteristics described by Maximillian—emerging from

the water, dancing and singing—describe the springtime behavior of these birds. Orioles are in the same family (Icteriidae) as blackbirds, hence the original classification of the bird as an oriole.

Wilson's Snipe

As stated in Chapter 3, the snipe is contentiously identified in two major Blackfoot bird sources, McClintock (1999) and Schaeffer (1950). Schaeffer goes into more detail in his unpublished notes of bird lore for the Blackfoot:

> The Snipe is never seen on the ground. Only when it undertakes its peculiar nuptial flight do the Blackfeet get sight of it. It flies into the air, circles, dives toward the ground and gives its call. During its flight a cloud will appear towards the mountains. Lightning and thunder will be heard. Soon it will begin to rain. The bird continues its flight, as it is not afraid of the storm. For this reason it is called the "rain-maker" [Schaeffer and Schaeffer 1934: GM, 1100-144].

The snipe is also associated with the Blackfoot Beaver Bundle as one of three birds that are thought to bring rain. They may also have the ability to stop rain. The snipe's association with rain and storms is also reaffirmed in two Blackfoot songs for the snipe, which exclaim, "The rain drops are my medicine" and "The hail drops are my medicine" (Schaeffer and Schaeffer 1934: GM, M1110-144).

"Cold Bird"

The Snow Bunting (*Plectrophenax nivalis*; Plate 44), which Schaeffer refers to as the "winter bunting," is known to the Blackfoot as "cold bird" or "winter bird" (Schaeffer 1950:45). They believed that the Chestnut-collared Longspur (*Calcarius ornatus*) and the Snow Bunting were one and the same, the bunting exhibiting the winter plumage and the longspur, summer plumage. The changing color of the "cold bird" could signal a severe winter, and the bird also had the power to bring cold weather and snow. The "cold bird" also had the ability to trick people by taking the form of a human sitting next to a large fire during the winter (Schaeffer and Schaeffer 1934: GM, 1100-143). If someone came to sit at the fire, they would find it so hot that they would take their clothes off and in doing so, freeze to death.

Warfare

Bird qualities and emotions play a significant role in warfare, particularly in protecting and lending their sight and speed to warriors in the battlefield. Bird parts are also significant in warrior regalia and war bundles.

Thunderbirds and Eagles

Beyond its association with storms and rainfall, the eagle offers the greatest protection to warriors as they face the perils of battle against the enemy. As unparalleled avian predators, eagles are capable of taking down opponents of formidable size, such as deer, mountain goats, pronghorn, and cranes, with their speed, agility, and aggressive flight. These qualities translate into indispensable skills in battle, making eagles a powerful warrior ally. Golden Eagles (*Aquila chrysaetos*; Plate 18) are known as "war eagles" to the upper Missouri River tribes, and their feathers and other parts are symbolic of warriors, acts of war, or may signify a special bond with the eagle as one's spirit animal.

After his hailstorm experience, Lone-tree was visited periodically by eagles in dreams. He also received success in warfare from the Bald Eagle and went on to become a renowned war captain (Lowie 1922b:335). Plenty-Coups was adopted by the Golden Eagle who gave him the power to be a successful war leader, reaching notoriety on the warpath and becoming chief before he was 25 years old. He kept a yellow-dyed eagle feather with him at all times in order to keep the golden eagle's presence with him (Linderman 2002:173). Similarly, Wolf Chief was given an eagle feather by his father after he was nearly shot by an enemy in battle (Wilson 1911). His father hoped that if he treated the feather as sacred, it would protect Wolf Chief from danger.

Some go as far as to say that the only way to be successful on the warpath is through the power of a thunderbird such as an eagle or hawk (Rodnick 1938:46) (Figure 5.2). Eagles may also preemptively warn of approaching enemies to those who have the eagle as their medicine. Sore-tail, a Crow man who was adopted by an eagle and founded the Eagle Chapter of the Tobacco Society, kept a pet eagle who was known to whistle when enemies were approaching so that the Crow people could tie up their horses to prevent horse-theft (Lowie 1922b:364).

Celestial bodies are also recognized by the Crow as interconnected with the eagle as physical embodiment of a war guardian (McCleary 1997:35). One Crow consultant knew of a man who had a relationship with the Morning Star, the son of the Sun, as a spirit helper in war. The Morning Star responded to a man who went to fast, seeking help in war. The star would transform himself into an eagle when this man went into battle and would give him the power to fight and defeat his enemies.

Figure 5.2. A Golden Eagle head used as a war charm, c. 1903. American Museum of Natural History (AMNH) 50/4549A.

Eagles are still an integral part of spiritual and ceremonial life across the Plains, and eagle feathers are currently in use with the tribes when available (due to the restrictions for acquiring eagle feathers through the federal permit system) to protect and honor modern day warriors, those who have served in the military (Figure 5.3). Crow servicemen and women are often given "medicine feathers" of the eagle or another bird from their paternal clan uncles for success and protection during war (Voget 1995:29).

Crow historian and anthropologist Joe Medicine Crow joined the U.S. Army during World War II and served in

Figure 5.3. Arikara consultant holds two feathers received for her military service (2010). Photo by W. F. Murray.

the 103rd Infantry Division. Whenever he went into battle, he wore his war paint beneath his uniform with a sacred eagle feather beneath his helmet. During his service in World War II, Medicine Crow completed all four tasks required to become a Crow war chief: he (1) touched a living enemy soldier, (2) disarmed an enemy, (3) led a successful war party, and (4) stole an enemy horse. He is the last member of the Crow tribe to become a traditional war chief (http://www.custermuseum.org/ medicinecrow.htm).

Yellowtail, a spiritual leader for the Crow and chief of the Sun Dance ceremony, gifted an eagle feather once used in the Sun Dance to his adopted son before he left for combat in the Vietnam War (Fitzgerald 1991:74). The feather protected his son from being injured or killed while away at war, and Yellowtail likened it to the feathers carried as medicine by warriors in the past.

As Murray (2009, 2011) explains, similar practices continue today, with members of the community relying on knowledgeable religious leaders and experts to acquire eagle feathers through the National Eagle Repository system. Among the contemporary Mandan and Hidatsa, the Eagle War Staff can only be carried in ceremonial occasions by veterans who have taken an enemy. Members of the community may receive an eagle feather in recognition of outstanding acts and bravery in military combat.[1]

Hawks

Hawks are similarly associated with warfare. They are often called on by warriors who have received the hawk as their patron through prayer, dreams, and material objects such as stuffed hawks or feathers, which are carried with one into battle (Bowers 1992:231, 245, 254). Noted for their fleetness, the hawk gives warriors unprecedented speed on the battlefield, which allows them to outrun their enemies and steal many horses (Parks 1991:351–363). Speed could help warriors to count coup, an honor that came with certain acts of bravery such as being the first to strike an enemy in battle. The Small Hawk ceremony, performed by the Mandan in January, was meant to honor the hawk and ensure good fortune in future battles. The ceremony was also used for buffalo calling in the summer (Bowers 2004:108). The hawk is also linked with the swallow. Together their medicine helps ensure a successful war

party, and they are also known to have introduced horses to the Hidatsa (Bowers 1992:245, 435–436).

Independent from the skills relating to the hawk's physical prowess, the hawk is known for other powers not directly related to observable traits of hawks, such as making warriors impervious to enemy fire. Hawk feathers were worn by warriors who had not yet received a vision from a spirit animal to protect them. The feathers held defensive energy to shield warriors who were otherwise very vulnerable without a spirit guardian (Weitzner 1979:289). One Crow story gives the hawk similar credit for its protective qualities when the hawk gives a child the ability to fight the enemy in battle without being shot (Lowie 1918:223).

A Blackfoot consultant mentioned that the Rough-legged Hawk was his great-grandfather's war medicine: "The Rough-legged Hawk, I know my great-grandfather, that's what gave him power. It was war medicine. A lot of the birds provided spiritual protection and power to individuals when they had their vision quest. A lot of the individual birds would do that."

Prairie-chickens

One of the 12 Hidatsa clans is named for the prairie-chicken (Wilson 1908:118). When asked about the meaning behind the name, Wolf Chief, a member of this clan, suggested that the name had to do with all the brave men who had come from that band (Wilson 1908:119). Prairie-chickens can be quite aggressive in their communal lekking displays, which might be analogous to Hidatsa war party activity. Wolf Chief commented that "In fight they [were] always at front," and he continued in the voice of the Prairie-Chicken Clan, "Now you look at me! I belong to the Chicken band! In old times I fight in war. Now war is over but another war is on. I try to fight evil . . . so you see the Chicken spirit is not dead!" (Wilson 1908:119).

Ravens

The raven is best known in warfare for its "long vision," which he lends to men on the warpath to better see their enemies. Ravens may also warn of an enemy ambush in a variety of ways. According to the Crow, those with Raven medicine gained the ability to understand any bird when they asked Raven where to find the enemy (Nabokov 1970:151). Blackfoot warriors looked for Raven when out on war expeditions. If ravens were spotted on the trail ahead and two had their heads close together like they were whispering it was a warning of an approaching enemy (McClintock 1999:483). Blackfoot men were

---

[1] The use of eagle feathers by women warriors may be a practice restricted to the contemporary upper Missouri tribes. Traditions of the Santee Sioux who now live along the lower Missouri River hold that only men wore or owned eagle feathers in the past while women wore hawk feathers (Bethke et al. 2014).

known to tie the head and neck of a raven to the back of their head. The raven hairpiece would alert a man if he was coming upon an enemy and did not realize it by tapping its beak on the back of his head (Grinnell 1920:261). A Blackfoot consultant further explained where the raven hairpiece gets its power and who has the right to wear it: "The crow or raven is a hairpiece, the whole body, and the *Beaver Bundle* gives it to the person, the owner, it lends that hairpiece out. They tie it to their hair. And as they are riding through enemy country, that bird pecks their head. It helps them to be alert, finding the enemies."

Blackbirds

Blackbirds can often be observed on farmlands, and it is not uncommon to see them near livestock and horses. Young boys would shoot and eat the numerous blackbirds that lingered around horses to pass the time as they minded the herds (Wilson 1924:155). Because of the blackbird's close relationship with horses, those with blackbird medicine would lead war parties and horse raids because the blackbird could always find the enemy's horses (Nabokov 1970:152). Those who had medicine from the blackbird were always chosen as leaders of a Crow war party, regardless of whether it was a young boy or an older man because, despite one's level of experience, the blackbird would help find the right place unerringly (Lowie 1922b:362; Voget 1995:xvii). Blackbirds may also be associated with successful horse raids and scouting because they are known for their "power to fool" (Wissler and Duvall 1908:114), a valuable asset for acquiring horses without any trouble from their owners.

Cranes

The crane sometimes has associations with war, and it is referenced by several sources as a vicious-natured bird that is prone to attack, according to the Scarface story of the Blackfoot (Grinnell 1920:100; Wissler and Duvall 1908:63).[1] Medicine Crow, one of Lowie's key informants, received a vision of a crane after going out onto the prairie to mourn and fast for a friend who had been killed recently in battle (Lowie 1919:164, 1922b:341). The crane appeared to Medicine Crow with a fresh scalp around its neck, and near the bird was a bush full of ripe cherries. He believed

---

[1] Grinnell's account (1920:100) does not refer specifically to the crane, but to "great birds which have long sharp bills." In the account recorded by Wissler and Duvall there are two encounters with vicious birds, first geese and then cranes. The "long sharp bills" of Grinnell's account suggest cranes rather than geese.

that he would avenge his friend's death during the season when the cherries were as ripe as his vision, and indeed he killed a Dakota warrior during that season. He also bought crane power from another man to strengthen the power he already had, and went on to become a leader in war (incidentally, so did his son). The Blackfoot word for the crane, *sikam*, is used also in the phrase "running crane" or *sikamokskatsi* to signify the way that a war scout runs in a zigzag pattern like a crane (Blackfoot consultant, 2011).

Doves were reported to be one of the most difficult birds to fell with a bow and arrow, because they usually required two shots to kill (Weitzner 1979:198). Perhaps for this reason, doves (likely including Mourning Doves [*Zenaida macroura*] and Rock Doves [*Columba livia*]) are associated with safety and the ability to escape danger, offering protection from the enemy during battle in a manner similar to the hawk. It is interesting to note how Native Americans incorporated Rock Doves into their views. This Old World species was introduced to North America by early European settlers. Before going out on a raid, the war party might be blessed using dove medicine to ensure their safe return (Hungry-Wolf 2006:139). The dove may also be associated with a "safe return" because it was the dove that guided the Mandan to land after a great flood. The Dove or Pigeon Society of the Blackfoot is connected to war because of the dove's protective abilities. The society was given to an old man who was mourning his son's death on the warpath; the doves gave him the dances and songs for the society and helped him to avenge his son's death (Wissler and Duvall 1908:105). Warriors participate in the Dove dance by imitating their approach to the enemy with their bows and arrows (Hungry-Wolf 2006).

Robins

The American Robin (*Turdus migratorius*; Plate 43) is not noted for its bravery or speed like many other birds associated with warfare but instead is a harbinger of peacefulness and security (Hungry-Wolf 2006:136; Schaeffer 1950:43). When its familiar song was heard around camp, it indicated that there were no enemies nearby. It was a valuable ally in the old days of warfare, and its departure each fall was lamented.

Meadowlarks

As with the robin, the presence of meadowlarks (Plate 46) around camp is taken as a sign of peace (Hungry-Wolf 2006:136). There are many stories where a meadowlark warns people of danger or attack so that they are able to escape in time (Beckwith 1937; Bowers 2004; Parks 1991;

Parks et al. 1978). The meadowlark is almost always correct about predicting danger and attack, as those who do not heed its warning find out. In Mandan traditional history, Speckled Arrow disregarded the meadowlark's advice to fortify his village with a palisade and watched it be destroyed as nefarious enemies lit his village on fire and everyone was killed (Parks et al. 1978:85–89). Archaeologists have since uncovered bastions and palisades around populous Mandan villages, such as Double Ditch (Kvamme and Ahler 2007; Will and Spinden 1906).

Chickadees

The chickadee (*Poecile* sp.; Plate 40) is small but wise and a great negotiator of peace, although he can also hold his ground against the enemy in battle (Linderman 2002:37–38, 41). The chickadee carries an almost immediate association with the famous Crow warrior and chief, Plenty Coups. For example, as a chickadee flew by during an interview with a Crow about the cultural resources in the vicinity of Fort Union Trading Post, the consultant commented on the Plenty Coups' war power, gained from the chickadee (Zedeño et al. 2006:224). The chickadee was Plenty Coup's medicine, giving him great power and success in battle. He attached the stuffed skin of a chickadee to his hair before going into battle in order to keep the power of the chickadee close at hand (Linderman 2002:79–80).

Woodpeckers

Woodpeckers were valued as war medicine by Blackfoot warriors. A Blackfoot consultant explained: "Woodpeckers are a favorite war medicine. They seem to fly from tree to tree, dodging and dodging. They are hard to shoot. . . . The warriors use them as their medicine because they are so hard to find."

Magpies

Magpies are sometimes used as medicine objects in order to "take the lead in war" or for success in capturing horses on raids (Lowie 1922b:392, 419). Magpies also have awesome and coveted powers in Blackfoot tradition.

Hunting

Hunting was and is an essential part of Plains Indian life. Big game—especially bison—was a crucial component of their diet. The Blackfoot, for example, regarded bison as *nitápi waksin* or "real food" and other edibles as *kistapi wksin* or "nothing food" (Schultz 1962:30). Important meat supplements included berries and wild vegetables and tubers (Rodnick 1938:27). Hunting was also closely tied to the economic, social, and spiritual orders. Fluctuations in the hunt from year to year and season to season, with intervals of surplus and scarcity, required more than just hunting skill. The acquisition of a hunting spirit, as well as participation in game-calling and buffalo-calling ceremonies like the Assiniboine Home-Building Dance or *Woti'jaxa* or the Mandan *Okipa*, helped to overcome the risks faced in the uncertainties of food supply. The Blackfoot Sun Dance, better known as *Okan* or Medicine Lodge, includes a magical performance that imitates the bison drive. The Blackfoot also had bison-calling rituals. Bundles associated with them include the Beaver Bundle and the *Iniskim* Bundle of the Black and Yellow Buffalo tipis (Zedeño 2013).[1]

Thunderbirds and Eagles

Eagles are strongly associated with hunting because they are expert predators, capable of catching rather large fish, reptiles, and mammals, and known for their swift and deft maneuvering in pursuit of their prey. Many hunters call on eagles for help to ensure a successful hunting trip, asking for the eagle to share some of its hunting skill in return for an offering or prayer, or employing an eagle medicine object to aid them in the hunt.

The Assiniboine hunter known as Returning Hunter made sure to leave the first killed deer of the season for all the birds, but especially for eagles and the "bullet hawk" (this bird could not be further identified) (Rodnick 1938:25–26). This act would make or break a successful hunting season, as the birds may be offended without such an offering and would bring bad luck to a hunter who disrespected the custom. During a particularly fruitless hunting season, Returning Hunter desperately prayed to the Thunderbird for assistance and was answered as a huge eagle swooped down and seized a pronghorn, killing it and leaving the body for him. Returning Hunter offered the liver of the pronghorn to the eagle, and packed the rest of the meat to take home to his family (Rodnick 1938:46). Eagle and hawk spirits came to him in visions when he was fasting and helped him to hunt in subsequent expeditions (Rodnick 1938:26).

In an Arikara story, a man is pitted against his potential father-in-law, who challenges him to complete a series of seemingly impossible tasks in order to marry his daughter.

---

[1] Marine fossils known as *Iniskim* gave themselves to a woman who was gathering wood during a time of scarcity; the woman went on to tell the hunters how to call bison (Reeves 1993). The bison-shaped Iniskim were essential in bison-calling, tobacco planting, and war powers.

The father-in-law tells the young man to go out and kill a white elk. Although the eagle had never seen one before, he assisted the young man in the hunt and helped him to find and kill the rare white elk (Parks 1996:197–206).

In several of the major oral traditions among Missouri River tribes, Thunderbirds in the form of eagles implore humans for help in hunting and defeating an enemy (see Beckwith 1937:92–95; Dorsey 1904:78; Parks 1996:206–209). This is especially common in stories involving culture heroes, and exhibits the fluidity of relationships and power between the spirit world and human world. Spirit animals are regarded for their power, yet at times they require help from others to complete the most important tasks. Reciprocity, rather than hierarchy, mediates relationships among different things and beings in the universe. This contrasts sharply with ancient Old World narratives, such as Greek and Roman myths and the Bible, in which the gods often help mortals but the relationship is not reciprocal, and neither is the relationship between people and animals (except perhaps for the Phoenix, Serpent, or Unicorn).

Hawks

An Assiniboine story recounts a trial of hunting skills between two hawks, Red Hawk and Black Hawk, for jurisdiction over the fate of a group of starving people (Lowie 1909:201). Before people knew how to hunt, they were going hungry. Red Hawk was watching them, waiting to eat them up, when Black Hawk saw what was happening and wanted to help the people. A hunting competition ensued between Red Hawk and Black Hawk for the fate of the starving people. Red Hawk challenged Black Hawk, who normally hunted mice, to catch and kill a rabbit in the woods. With some difficultly, Black Hawk eventually caught and killed his first rabbit. Then Red Hawk attempted to catch mice in the forest, but failed because his wings struck the trees and scared away the mice. When Black Hawk proved that he could catch the mice with his noiseless flight, Red Hawk was forced to leave the people alone. Black Hawk then taught the people how to hunt so that they would no longer starve.

Not only could the power associated with a certain animal be imparted through a vision or dream, but animal powers might also be acquired through purchase from a spiritual leader or medicine man. In addition to his visions, Returning Hunter received power from the hawk by paying a medicine man (*wincasta wakon*), who gave him the power to wear two hawk feathers on his head so "that he might have good luck by imitating their way in hunting" (Rodnick 1938:26). Zedeño (2008a:365) explains that medicine bundles are "transferable through ritual, purchase, or inheritance." The ceremonial purchase of a medicine bundle or object transfers the powers and abilities associated with that object to the new owner.

Chickadee

Returning Hunter used a chickadee charm he acquired from a holy man to gain better luck in hunting (Rodnick 1938:26). While out on a hunting expedition near the Missouri River south of the Fort Belknap Reservation, he spotted a black-tailed buck a long way off and two chickadees nearby. He observed the chickadees alight on the antlers of the deer. Soon after that the buck fell into a deep sleep. This allowed Returning Hunter to approach the deer and shoot it before it awoke.

Ravens

The raven is always able to find food (Grinnell 1920:101). Ravens share this ability with hunters in several ways. In addition to their keen sight, which is invaluable for hunting, ravens are augural for finding game and even leading hunters toward the animals. If two ravens were spotted playing on a ridge during a fruitless buffalo hunt, it was a signal that the hunting party should move in the direction where the ravens were playing and they would soon find buffalo (McClintock 1999:483). A raven will also call to a hunter (human or animal) when game is near, flying back and forth between the game and the hunter until the hunter heeds its gestures (Grinnell 1920:261).

Magpies

Magpies are often characterized as mischievous birds because they steal scraps of meat drying on outside racks (Bowers 1992:180). These birds are also looked upon as friends because they are one of the first birds to show up at a baited eagle-trapping pit, and they are thought to attract eagles to the bait (Gilmore 1929:30; McClintock 1999:481–481). When the magpies scattered away from the pit, it was a sign that an eagle was approaching (Wilson 1928:202). A Blackfoot man named Spider received power from the magpie in return for feeding its young, and the magpie showed him where the best place was to trap eagles (Wissler and Duvall 1908:137). Each year he returned to this place, fed the magpie's children, and had success in trapping eagles. He later transferred this practice to another man as payment for a medicine pipe, explaining that the man must feed the magpie's children in order to receive their help in eagle trapping.

Planting and Harvesting

Old-Woman-Who-Never-Dies is the culture hero who brought the harvest to the Mandan and Hidatsa, helping the crops to grow by bringing rain and sending birds from her home in the south as her messengers. She is associated with maize, the main crop of the tribes, but also with beans, squash, and other agricultural produce. As her messengers, migratory waterfowl and other birds take on a significant role in the ideology of agriculture and agricultural ceremonies, such as the Old-Woman-Who-Never-Dies rites, which reenact important events of the past, and the ceremonies of the Goose Society (Bowers 2004:344). The Old-Woman-Who-Never-Dies bundles are associated with corn, beans, pumpkins, sunflowers, blackbirds, geese, ducks, and cranes, as well as elk, deer, bears, and dogs.

Blackbirds

Blackbirds are the caretakers and "helpers" of Old-Woman-Who-Never-Dies, watching over the gardens (Beckwith 1937:53; Bowers 1992:334). Blackbird heads are included in the Old-Woman-Who-Never-Dies bundles to signify their status as special helpers who eat the pests from the gardens (Bowers 2004:184–185; Peters 1995:112). The term "blackbird" probably refers to a host of different species that generally resemble one another, such as the Brewer's Blackbird (*Euphagus cyanocephalus*), the Common Grackle (*Quiscalus quiscula*), and the Brown-headed Cowbird (*Molothrus ater*), among others who live in the Missouri River Basin.

Oral traditions relate the sometimes conflicted or misunderstood relationship that blackbirds have with crops as both an aid and a nuisance. Their diet of insects like weevils, cutworms, termites, grasshoppers, and caterpillars can help reduce the number of destructive pests in the garden, but blackbirds are also known to munch on those same seeds and grains themselves later in the growing season when insects are less plentiful (CLO 2011). In one widespread story, a holy old woman adopted a young boy who had fallen from the sky, born of a woman and a star (Bowers 1992; Bowers 2004; Parks 1991). The boy grew up watching his grandmother parch the corn in her lodge and feed it to the blackbirds while she went out to the garden. One day he decided to mimic his grandmother; he parched the corn and called the blackbirds into the lodge, but then he trapped them inside the lodge and killed them to be eaten later. Not understanding the relationship between Old-Woman and the blackbirds, he told his grandmother that the birds would no longer eat up all of her corn. The old woman mourned the birds who would no longer watch over her garden, so she gathered their bodies and took them into the woods, where she brought them back to life.

Cranes

Because they carry the summer with them as they migrate, controlling the seasonal changes in the weather and therefore the planting and harvesting seasons, cranes are very closely tied to agricultural rites (MWP 1942:30). The crane is also one of the messengers of Old-Woman-Who-Never-Dies, and it is believed that in their southward migration they travel to her home on an island, where they live alongside many other sacred birds and animals, as well as corn and garden spirits (Bowers 2004:245).

Not only is the crane associated with the ceremonies for growing corn and other produce, but it is also linked to the harvest of a sacred root. Several Blackfoot stories relate how a crane came to a young woman to show her how to dig the "sacred root," sometimes referred to more specifically as a "sacred turnip" (Wissler and Duvall 1908:59). A woman who lived in the sky with her husband, Morning Star, was warned by the Moon not to pick a large turnip; however, she became curious about it one day and tried to dig the root (McClintock 1999:494). When her digging stick became stuck in the ground, Crane showed the woman how to use it correctly.[1] Crane also taught her certain songs to sing as she dug and picked the turnip. In another story about the Natoas Medicine Bonnet, a woman was visited by an elk who called on many other animals to give her power; the crane offered the use of his bill to dig the sacred turnip for a woman to use as medicine (Wissler and Duvall 1908:83–84). The sacred turnip digger is always tied to the "Medicine Case" of the Blackfoot, which contains the Medicine Bonnet given to the woman by the elk (McClintock 1999:500).

Geese, Swans, and Other Waterfowl

Geese are another of Old-Woman-Who-Never-Dies' helpers, and are very closely tied to the planting and harvesting of maize (Beckwith 1937; Bowers 2004; Lowie 1915a; Peters 1995; Will and Hyde 1964;). Sometimes geese are viewed as representations of the maize, while at other times they are seen to be one and the same with the corn spirits (Will and Hyde 1964). The Goose Society of the Mandan is associated with agriculture and rainmaking. When a member of the Goose Society dreamt of geese during the winter,

---

[1] In some versions the crane offers its beak for the woman in the sky to dig out the sacred turnip.

they would hold a feast for the geese and pray for a good harvest (Bowers 2004:202–203). In the past, the Mandan and Hidatsa would leave offerings of dried meat at the entrance of the village for Old-Woman's messengers to carry back to her as gifts until this practice was outlawed by the federal government (Will and Hyde 1964:271). These offerings and ceremonies were made to ensure that the geese returned with the corn each year.

Similarly, the swan is also a messenger for Old-Woman-Who-Never-Dies, sent in the spring to symbolize the gourd, and the duck represents beans (Beckwith 1937:53; Peters 1995:113). Other waterfowl (for example: coots, pelicans, kingfishers, ducks, loons, cranes, and grebes) are also considered her helpers and are more generally associated with agriculture.

Bittern

The bittern's association with the sun and water give it an important role in the Blackfoot tobacco rites (Hungry-Wolf 2006:138). A bittern skin was included in the Beaver Bundle that was opened during the Tobacco Planting Ceremony, although it has since been replaced with a swan skin after its mysterious disappearance. During the planting ceremony the bittern's actions are imitated by the bundle-holder's wife, who gazes up at the sun like the bittern is thought to do. The bittern assures that there is enough water for growing the tobacco plant through its association as a water bird. Blackfoot consultants observed that sacred tobacco planting has been revitalized owing to the recovery of requisite songs that were recorded and archived by Frances Densmore in the early twentieth century.

Snow Bunting

Tobacco planting rites of the Blackfoot, associated with the Beaver Bundle, are also linked with the "snowbird" (Wissler and Duvall 1908:79–80).[1] In the past, several men owned a Beaver Bundle communally. One man had to leave for a journey, but asked the other men to hold off on planting tobacco until he returned. They did not wait for him, but the pronghorn and snowbird offered the man their dung to fertilize his tobacco plants. Several

birds (unidentified varieties) agreed to help him obtain some tobacco seed from the Sun. Only one bird made it to the Sun and returned with tobacco seed, which the animals then proceeded to plant for the man while they taught him new tobacco songs. These plants were the only tobacco to grow during this season. The plots that the other men planted were trampled by buffalo. The man shared the tobacco that he had grown with the help of the snow-bird and other birds and animals, and the songs he learned were incorporated into the Beaver Bundle for tobacco growing.

Guards and Protectors

Birds are sometimes designated as guardians or protectors of supernatural beings, humans, or resources of medicinal and spiritual significance such as a certain location, plant, or animal. Their songs and calls can act as an alarm for trouble approaching and their flight allows them to quickly deliver messages, but they may also possess extrasensory abilities to help or hinder those who seek whatever they guard.

Thunderbirds and Eagles

Eagles also protect important cultural resources for tribes around the Missouri River. Eagles guard spiritual resources, only guiding those whom they recognize as having the right to use those resources to the correct location and deterring those who lack permission to use them. Yellowtail recounted his first time going to collect deer-weed and lightning root (species unknown) with another Crow friend. They prepared an offering of tobacco and began to pray before collecting their medicine plants, and two eagles alighted nearby, watching them. They prayed:

> Thank you, Aho! You are here to guard this sacred medicine and this is good. It is to be preserved for the benefit of those who are authorized by our Medicine Fathers to use it in their healing. We have been authorized to use it, and we are here to take just enough to help us heal the sick and needy. Your power is great, and you know what is in our hearts. Help us, show us the medicine. Aho! [Fitzgerald 1991:47].

They went on to collect the plants, as the eagles remained perched on a rock looking on. Yellowtail passed on the medicine rights to his grandson, who was also successful in finding the roots when he went out to collect them. However, Yellowtail also noted that several people who did not have the right to collect those plants were unsuccessful in finding them, wandering around for

---

[1] The snowbird is referenced in the literature of several tribes, and was identified by Maximilian as a Snow Bunting (*Plectrophenax nivalis*; Plate 44) (Weitzner 1979:198). A Hidatsa story describes the snowbird as "whitish birds with black streaks on their wings" (Beckwith 1937:104). Based on this description and the territorial range of birds matching this description, Snowbird could also refer to the Loggerhead Shrike (*Lanius ludovicianus*) or Northern Shrike (*Lanius excubitor*).

hours, searching to no avail. He attributed their failure to the eagle guardians (Fitzgerald 1991:48).

Yellowtail also regarded his medicine father, the Golden Eagle, as a guardian to him at all times (Fitzgerald 1991:220). Fitzgerald, who chronicled the life history of Yellowtail, recalled the omnipresence of the eagle with Yellowtail even as he traveled. They spotted an eagle when Yellowtail first arrived at the author's home in Indiana, and on other trips they took together. Yellowtail was always sure to thank the eagle as his guardian for his safe travels and looking after him.

Cranes

Both the Sandhill Crane and the Whooping Crane (*Grus americana*; Plate 12) were common visitors to the Missouri River around the time of the first arrival of Europeans to the area, although whooping crane populations are now much reduced due to habitat destruction and hunting. Cranes have an impressively loud, bugling call that makes them well-suited as guards because they can quickly sound an alarm if someone is approaching. In addition, Sandhill and Whooping Cranes are among the largest birds in North America and can be quite aggressive at times, making them an intimidating prospect as guards in these stories.

The Snow Owl Ceremony of the Mandan originates in part from a story about a man who fell asleep while eagle trapping and woke up in a strange place (Bowers 2004:287; Lowie 1918:141–143; McCleary 1997:65–69). He was challenged by an arrow maker to kill a certain elk that was protected by two cranes. With the help of ants (or moles, depending on the story teller), the man was able to tunnel under the elk without the cranes realizing until it was too late. Later in the story, the man must kill Red-Hair, a powerful man who lived very far away in a lodge guarded by a crane and a coyote who bugled and howled when anyone approached. Again, the ants helped him in his plan to kill Red-Hair, transforming him into an ant so that he could pass by the crane and coyote as they napped at noon. The crane awoke, but it was too late and the man was able to sneak by in disguise and kill Red-Hair.

Travel

Large migratory birds such as the crane, goose, and swan are often associated with narratives of travel because of their large size and flight pattern, which make them visible while en route to a new locale. People may ride on the backs of these birds in flight as part of an epic journey, and swans and geese also help travelers to cross large bodies of water. The Crane Chief carries a Crow man on his back to the land of the birds where the Crow teaches cranes to eat meat. The man collects eagle feathers to make two war-bonnets, and receives power from a crane, a hawk, a condor, and the Crane Chief (Lowie 1918:158–161).

The Assiniboine and Crow both have oral traditions in which the trickster character asks to travel on the backs of a group of geese preparing to take flight (Lowie 1909:108, 1918:38). They explain that flying is difficult, but agree to take him along on the trip. Eventually the geese fly above a camp. The people below recognize the trickster in goose form and he falls out of the sky, humiliated. People may also transform into geese in order for their spirits to travel in the afterlife, as one Blackfoot consultant explained: "When warriors fall in battle, they turn into geese and come back home. Their spirits come back as geese and you can see their chiefs forming the tip of an 'A' that's how you know them."

The swan is most often mentioned as a helper to travelers over long distances or through tough terrain. In fact, in 2011 a Blackfoot consultant noted that the Blackfoot word for the swan is *tsekomkhkayii*, meaning "white going home," referring to their status as travelers (see also Schaeffer 1950:40). A swan and an otter help a man who was accused of adultery and deserted on an island to return safely to his home, after which he adopted the otter skin and swan skin for his medicine (Wissler and Duvall 1908:98–99). Two swans help Scarface cross a large body of water that seems impassable on his journey to the Sun's lodge to get his permission to marry his love, who had promised herself to the Sun (Grinnell 1920:95–103). In another story, two swans also help a group of children to flee from an old cannibal witch by forming a bridge over the river with their necks. After the witch offends the swans, however, they trick her into crossing their necks only to throw her off midway across and she drowns (Lowie 1909:142–144).

Predicting the Future

Some people among the Blackfoot believe that birds have the ability to foresee the future. They communicate their predictions by pecking images onto rock faces. The peckings are far too high to have been reached and created by humans, so they are attributed to birds.[1] The bluebird

---

[1] Schaeffer adds that James White Calf believed that the spirits make the peckings. To paraphrase James White Calf, he believed that the spirits know all things, and this is how they can produce likenesses of any individual and predict their future (Schaeffer and Schaeffer 1934: GM, Blackfeet M-1100-143).

(*Sialia* sp.; Plates 41, 42)[1] has extraordinary power in creating these images in comparison with other birds. Short Face, one of Schaeffer's consultants, explained that Thunder created the bluebird and "gave the bird its paint" (Schaeffer and Schaeffer 1934: GM, M-1100-143).[2] The bluebird could peck images into the cliff that showed up in many colors, unlike the other birds who made plain images. Short Face told Claude Schaeffer that long ago, a person named Many Spotted Horses sought power by fasting at a certain cliff in Blackfoot country. A spirit warned him to leave the place, as humans are forbidden to see the birds writing on the cliff. The novice continued to fast there, despite the warning that he would not live long if he stayed. That night he saw birds of different colors flying about the cliff. The bluebird said that his blue color would last longer than others. He awakened to see blue birds and others pecking out figures on the stone cliff. All the birds pecked out plain drawings, but the bluebird was able to make drawings of different colors. The novice started to make fun of the birds, refusing to believe what they said. They warned him that he would not live long but would see his picture on the cliff. He awakened the next morning, went to the cliff and saw the representation of himself, with his head cut off. He started home. En route the Cree killed him and cut off his head. People still journey to the cliff to see what the future has in store for them.

According to Blackfoot consultants, this story probably refers to the site known as Writing-on-Stone in Alberta. War parties would pass by these cliffs on their way to battle to see their fates, even though there was no known way one could change the fate inscribed on the stone. They also note that sometimes the bluebird itself can represent Thunder.

Health and Healing

Cranes have the power to heal. A powerful holy man known as Cherry Necklace received power from the crane to remove arrowheads and bullets from people who had been struck in war when he fasted at Eagle's Nest Hill near Washburn, North Dakota (Bowers 2004:178). Using two dried crane heads and a stuffed crane as his medicine, the holy man sang the crane's song, bringing the stuffed crane

to life so that it could remove the bullet or arrow with its bill.

According to the experience of a Hidatsa man, meadowlarks have healing qualities (Weitzner 1979:197). Meadowlarks are rarely shot because the Hidatsa believed that the meadowlark, being such a loquacious bird, would scold them and call them bad names (Wilson 1911:100). They are said to cure deafness and muteness if eaten:

> When anyone in our village grew deaf so that he could neither hear nor speak, his elder brother or someone else went out and killed a meadow lark for the sick man to eat. We kept feeding deaf and dumb persons on meadow larks until they got better and could talk. We thought the lark was such a talker it would make the sick man talk.
>
> There was a man named White Face who believed this old story about larks. He had a brother, a boy named Bears-on-flat. Bears-on-flat heard a little but could not talk. White Face took this boy out hunting with him and they brought in many larks which the boy ate, and he got well and talked [Wilson 1911:100].

Similarly, the Cooper's Hawk (*Accipiter cooperii*; *omaxtsístsipanikim* in Blackfoot; Plate 20) has the power to cure blindness. The Blackfoot told Claude Schaeffer a story about a young man who had been blinded by the swelling in his face that resulted from an arrowhead wound to his nose and upper lip. A medicine man brought in an owl and a crow to try to heal the boy, but neither succeeded. "Then the hawk flew in swiftly and alighted on the boy's chin, fanned him with its wings, and pecked at the arrow-point with its beak. Before long the arrow-point broke away and came loose. The swelling which had blinded the boy was soon reduced" (Schaeffer and Schaeffer 1934: GM, Blackfeet M-1100-143).

Horse Medicine

In the Crow tradition, birds are associated with horse medicine, which could help a person succeed in horse raids and become horse-wealthy. One Crow man who had failed to receive a vision quest through fasting received horse medicine in a night dream (Lowie 1922b:327). In his dream he saw a bear and a horse accompanied by a bird (species unspecified). The Crow man heard this bird singing, and then he saw a man driving a large herd of horses with the bird tied to his head as medicine. This brought him wealth in horses. A Crow consultant identified the Northern Flicker (both morphs) as horse medicine:

---

[1] It is unclear whether this designation is simply referencing the color of the bird or the actual species of bluebird—Mountain (*Sialia currucoides*; Plate 41) or Eastern (*Sialia sialis*; Plate 42). If referencing the color, the blue jay (*Cyanocitta cristata*) is another possible candidate.

[2] Blue or green paint is used in the Blue Thunder Tipi and certain pipe and society bundles to represent Thunder (McClintock 1999; Scriver 1990).

Figure 5.4. Crow consultant holds a feather from the yellow-shafted Northern Flicker (2012). Photo by S. Clements.

I had a grandfather, and he had horse medicine. He had a brass bell tied around the red [flicker] feathers and horse hair, and he would hang it around the horse's neck. So that's what he always used it for, my grandpa, got flicker medicine. I don't know where you get it, but I know that he made that medicine for my late brother.

He further explained that while his grandfather used the feathers of the red-shafted flicker, feathers from both morphs of the flicker could be used (Figure 5.4). The Crow consultant also mentioned that the Red-winged Blackbird was important for horse medicine.

Blackbirds and magpies are also associated with horse medicine. The blackbird has a close relationship with horses, and those with blackbird medicine would lead war parties and horse raids because the blackbird could always find the enemy's horses (Nabokov 1970:152). Magpies are sometimes used as medicine objects for success in capturing horses on raids (Lowie 1922b:392, 419).

Ability to Scare

When Coyote was naming the Crow clans, he asked the prairie-chicken for its name, to which he replied, "I scare animals" (Zedeño et al. 2006:201). The prairie-chicken

has the ability to "scare anything that lives," because of its abrupt and raucous flight (Wissler and Duvall 1908:114). As it takes flight, the prairie-chicken noisily flaps its wings in a quick burst of energy before shortly sailing back to the ground. This commotion can be quite startling, and makes its way into several stories as a way to scare pesky tricksters or help a character in need. Sitcon'ski, the Assiniboine trickster, irritates a prairie-chicken hen so much that she takes off, startling him and throwing him off balance so that he falls into a lake (Lowie 1909:110–111). After several other animals trick Sitcon'ski, he becomes angry and convinces Thunder to bring a flood that kills all the animals that cannot fly.

Owl Powers

Many positive and negative roles are assigned to owls in the Missouri River region, so a separate discussion is warranted here. Major powers and abilities taken on by the owl include healing and health, as well as evil and death. The owl is also associated with wisdom, warfare, witchcraft, fortune-telling, material wealth, and hunting. The most widespread belief about owls is that they are an omen for death or very bad luck. The presence of an owl, the sound of its call, or certain phenomena associated with owl feathers and other parts may all signal imminent misfortune. Some pipes used in the Spring Medicine Pipe ceremonies of the Blackfoot are adorned with owl feathers. If one of these feathers falls off the pipe as the carrier dances with it, it is a sign that the dancer will not live very long (Hungry-Wolf 2006:139). The Blackfoot believe that an owl hanging around one's home and making a lot of noise is a ghost. If the owl cannot be shooed away, it will be shot.

A lot of people in my family didn't like the owl because of what they, you know, interpret, of bringing bad news to families. Just like when you hear a dog cry, you know you are going to hear bad news from somewhere. Something's going to happen. And I find that to be true to this day, because there were situations that have happened in our family that the dog and the owl had brought with them.

—Mandan-Hidatsa-Arikara consultant

Similarly, Agnes Yellowtail Deernose explained that many Crow believe that if an owl hoots twice it is bad luck, and further, that owls are spirits of the dead:

You can see why we kids were afraid of Red Woman and the dark. We also were afraid of owls, and I'm still scared of owls. When an owl comes and hoots twice, you will have bad luck. That happened to me and Donnie, my second husband. He had a little girl by his first wife. The girl was sick when this owl came and hung around our house. I didn't like that, for every evening he would hoot twice, and one evening the owl even sat on our roof. That is a real bad sign. I told Donnie to scare it away. Then he hooted twice and flew away. The little girl died soon after. You know, owls change themselves into a person or a baby crying, or a dog barking, and they imitate you. Sometimes they break twigs and throw them at you. That's what happened to Donnie. This owl flew alongside him and threw a big twig at him. I remember, too, that an owl hooted just before my dad passed away. That's why I am scared of owls and believe that they are bad luck. Owls are ghosts who didn't join the Other Side People and who hang around places where people are buried [Voget 1995:81–82].

The Crow further believe that even when ghosts do not take the physical embodiment of an owl, they linger around their burials and hoot like owls to communicate their presence (Lowie 1922b:380–381). The Mandan hold similar beliefs about the owl as an evil presence. In the Mandan Okipa ceremony, a man impersonates an evil spirit in the form of an owl. He frightens the dancers and the women and children in the lodge throughout the ceremony, simulating sex with the various dancers before he is finally chased out of the lodge at the end of the dance (Catlin 1967:85).

Hidatsa children regard the owl as a fearsome creature, and parents used this fear to keep their children on their best behavior by threatening that the owl would visit them if they misbehaved (Gilman and Schneider 1987:126; Weitzner 1979:221). Adults, however, better understand the complex but mostly beneficent nature of owls. Although Hidatsa children are afraid of the owl, long ago the Hidatsa were instructed by the owl to give their young children a certain haircut that resembled the "horns" (feather tufts) of the Great Horned and Eastern Screech owls (*Bubo virginianus* and *Megascops asio* [Plate 29] respectively; Wilson 1911:191). This practice derived from a story about their culture hero, Grandson, who saved a starving village that was eating meat cursed and poisoned by a strange creature living in a tree (Weitzner 1979:221). Grandson found the creature and pulled its tail up onto its head, making two tufts, and calling it an owl. The Hidatsa children's haircut imitated the owl's appearance and was meant to ensure

their strength and health. A Mandan-Hidatsa consultant told a story that illustrated the conflicting views some people have about the nature of owls:

> A lot of Hidatsa don't like the owl because they think it's a bad messenger. Death is coming. But some people told me, it's not only that, maybe he's bringing you good news too, you don't know. I had a chance, that snow owl, a guy had made a feather cap for me. It was pure white snow. My mom, she was mad because she didn't want the feather cap. She said, "You're not supposed to have that." That's not our way. But I was trying to be different too . . .[anyways] my mom didn't want it, so I never took it. I wanted it, it was different. You don't see anybody with that. But I had to go find the right to wear that feather cap. I didn't want anybody saying hey, you don't have the right to wear that.

The modern Blackfoot do not think of owls in a strictly negative way. A consultant noted that they "are supposed to represent past medicine pipe owners that come back as them. In recent times, probably a Christian belief, these things are seen as messengers of death and bad luck—it is confusing for our people." Yet, when the authors once heard the screech of an owl, a Blackfoot elder assured them that "it had been taken care of so nothing would happen," thus intimating that there is indeed a way to neutralize spiritually any negative event foretold by an owl.

Owls as Healers

Owls can be powerful healers. According to an Arikara origin story, the owl helps Mother Corn and the people on their journey and gives the Arikara roots and herbs to be used in healing (Dorsey 1904:33). The owl itself is often a healer in oral traditions, and those who have power from the owl, especially medicine men, can wield this power to help heal individuals in the community. Deceased Medicine Pipe owners would linger as owls near their burials and would visit anyone camping nearby during the night. James White Calf knew of one such place near the Marias River, where on a camping trip he was visited by an owl. James offered the owl a pipe filled with tobacco (Figure 5.5). He further explained that when an owl is heard hooting in the woods it can be called on for help and good luck in the future.

In one Mandan story, an owl helped a man who was scalped by instructing him to apply grease to his head and cover it with the cotton of the timber-rope plant [species unidentified] and a soaked buffalo heart (Beckwith

Figure 5.5. Artist's rendition of James White Calf's experience with owls and medicine pipe owners. Drawing by M. D. Tibau, 2015.

1937:109). The man recovered within four nights and had grown skin and a new head of hair. In the story of the Compassionate Brother-in-Law an owl also doctored a Crow woman who was gravely injured and blinded by her husband (Lowie 1918:188–190, 1922b:331). Unique to this story, the doctor owl is specified as a woman, one of the few times that a bird takes a female gender in a story from the tribes in the region. The owl was called on by a deer to help heal the woman's eyes; the owl lent her eyes to the woman and she healed. A similar story is also present among the Blackfoot. They tell that an unfaithful Mandan woman was discovered by her husband, who led her into the forest and blinded her with an owl, leaving her there to die (Schaeffer and Schaeffer 1934: GM, Blackfeet, 1100-144). The woman was discovered by another owl, who recruited a Cooper's Hawk to heal her with its medicine.

Owls are also an important part of many healing ceremonies and are often patrons of medicine men. In his time with the Hidatsa, Maximilian observed that owls were considered to be medicine birds and were sometimes allowed to live inside their earthlodges (Weitzner 1979:221). Members of the Arikara Owl Society often called on the owl spirit and used parts of the owl in their healing ceremonies (Gilmore 1932:215).[1] An owl plume

was employed by the one acting as a doctor as a tool to drive the disease from a patient.

Owls also possess strong healing powers according to the Crow and the Sioux. When Yellowtail's son had a persistent case of whooping cough, they visited the Dakota medicine man Little Warrior for a *yuwipi* healing ceremony. Of all the healing spirits, Little Warrior recommended that they call on the white owl[2] as a medicine doctor, because he was "very good" and would almost undoubtedly heal the boy (Fitzgerald 1991:70).

We could hear Little Warrior call upon the Medicine Fathers to come forward, and then we heard wings flapping and we know that birds of some type were in the room even though no door or window had been opened. All four of the rattles were also flying around the room, and sparks were jumping off the rattles. I heard the "clip, clip, clip" sound that an owl makes by opening and closing its beak. I knew that sound, and I knew an owl was very near me. . . . I also heard the sound of a beak touching the bowl that Little Warrior had placed next to us on the ground. All this was spooky, but I knew it was part of the ceremony, and I was not afraid [Fitzgerald 1991:71].

During the ceremony, the owl had drawn many things out of the boy's throat and deposited the sickening agents into a bowl. By the next morning, his coughing had stopped and he had fully recovered. The *yuwipi* ceremony is a variation of the "Shaking Tent" healing ceremony known among Plains and Woodland tribes.

Wissler (1912:81) recorded a number of dreams from Blackfoot medicine men in which owls transferred healing power to the dreamer. For example, "[o]nce I was watching a woodpecker and another bird sitting on a tree. They said, 'Now, watch us and we shall give you power to cure disease.' So they taught me songs and how to use them when doctoring the sick. I have used these songs to cure many people." Another medicine man told him:

One night I dreamed that I was out in a large forest. The trees were very thick. Presently, I heard an owl singing in a tree. So I got up from my bed, looked around, but could see nothing. Now, the fourth time this happened I saw an owl sitting up in a tree nodding his head. This

---

[1] The Arikara Owl Society was composed of medicine men who derived their power from the owl.

[2] Possibly the Barn Owl (*Tyto alba*; Plate 26) or the Snowy Owl (*Bubo scandiacus*; Plate 28), which are the two local species with the lightest coloring.

owl sang a song four times. Then he came down and I went up to him; as I approached him, he seemed to be a man. Also, a tipi stood there. The owl invited me into the tipi. I went in and sat down. The owl on one side and I on the other. Then the owl sang the same song again four times. The words in the song were: "Where you sit is medicine." Now this owl gave me his power and this power enables me to cure people.

Owls as Warriors

The Snowy Owl and the Burrowing Owl hold particular significance in warfare and bravery in battle. The Snow Owl bundle of the Mandan is used in several ceremonies including a medicine feast for the owls, Holy Woman rites, buffalo-calling, and arrow making rites. The buffalo-calling and arrow making rites are tied to hunting; the feast for the owl was directly associated with war and success on the battle field (Bowers 2004:282–283). Items from the Snow Owl Bundle might be carried by warriors into battle to ensure success.

The Burrowing Owl, known variously as the "prairie dog owl" (Bowers 1992:241) and the "long-legged owl that lives with the prairie-dogs" (Linderman 1972:44), is known by the Hidatsa and the Crow as having power to give men great bravery in war. To those who had the burrowing owl as their medicine, the nearby flight of the owl was a sign that the individual should show "unusual bravery" in battle (Bowers 1992:241). Pretty Shield, a Crow medicine woman, recalled that her father's medicine was the "prairie owl," which made him a great warrior (Linderman 1972). Although he was small, much like the Burrowing Owl, he was known for his bravery as he led the Crow to a victorious battle with only his medicine, the skin of the burrowing owl, and a coup-stick as his protection.

Owls as Witches

There are accounts of both the Mandan and Hidatsa keeping owls as soothsayers in their lodges (Parmalee 1977a:629; Thwaites 1906:382).[1] The Arikara believe that owls are associated with the Wichita Indians to the south, who were known to practice witchcraft (Dorsey 1904:22). There is also an element of witchcraft employed with the

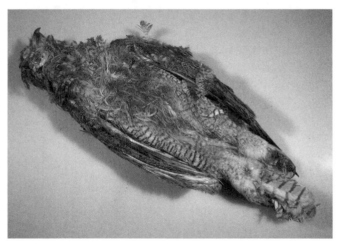

Figure 5.6. Great Horned Owl skins from the Blackfoot Medicine Pipe Bundle. AMNH 50/5717 (Date unknown).

acquisition of new members to the Blackfoot Medicine Pipe society (McClintock 1999:254). The society would call on the power of the owl and sing owl songs in order to capture the potential new member in a deep sleep like that of the owl during the day. They did this in order to cast a "spell" of sorts on the individual to prevent his or her escape (Figure 5.6). During the ceremony, the owl was thanked for his assistance through the offering of Siksocasim-root,[2] which was known as his favorite food.

---

[1] Maximilian identifies a "large grey owl, without doubt the *Strix virginiana*" as the type of owl kept by the Mandan for fortune-telling (Thwaites 1906:382). *Strix virginiana* is the Great Horned Owl (*Bubo virginianus*), which can range from brown to gray depending on its geographic location.

[2] *Siksocasim* is identified by McClintock (1999:485) as "Indian horehound," which is widespread in North America but not native. It may be another species altogether, but the *Siksocasim* is the holiest of plants and thus closely guarded. Horehound is a remedy for whooping cough; the strong connection between owls and horehound root may be linked to Little Warrior's suggestion (described previously) that Yellowtail called on the white owl to cure his son's whooping cough.

Owl Guidance

Owls occasionally act as guides. In the Arikara origin story, as Mother Corn and the Arikara people embarked on their journey, they came upon a dense and impassable forest (Dorsey 1904:15; Gilmore 1930:101). An owl (sometimes a screech-owl), is sent by the gods to help clear the forest, knocking down trees with its wings and creating a path to lead the people through. An owl also serves as a guide for a man trying to find his kidnapped sister in the Hidatsa story of Crow Necklace (Beckwith 1937:114–115). Crow Necklace comes across an old owl at a place called Timber Coulee, where the owl advises him the best route to take and ensures that he will be fed by the spirits, who will set traps to catch wild animals for him.

Owl Wisdom

In some cases, the owl has been recognized as an especially wise bird. In the Hidatsa story of the Two Twins, two twin boys are endowed with supernatural powers that enable them to defeat many monsters, including such figures as Long Arms and Pot-tilting Woman (Wilson 1908:1–45). After so much fighting, the Two Twins worried that more trouble might be headed their way, and they sought a companion who was even wiser than them. They met a small owl called *Hic-te*, who "was wise and knew everything in all the world," (Wilson 1908:43–44). This owl was able to keep them safe because he had all of the knowledge in the world and could foresee trouble. It is possible that in this story the owl plays multiple roles, as both a wise helper and a soothsayer.

CHAPTER SIX

Birds as Messengers

One of the more dominant roles of birds in Native American cosmologies is that of advisors or messengers, but this is not unique. Far into the past and around the world, birds have been recognized as skilled messengers, helpers and advisors by gods and men alike.

On coins found in Cyprus, belonging to the Union of Cypriote Towns and bearing the legend Κοινὸν Κυπρίων, appears the temple of Paphos, on which rest the holy doves of Aphrodite. Elsewhere, on sculpted monuments, they hover round goddesses; Astarte presses them to her bosom; priests and sacerdotal women carry them. They were encouraged to breed in sacred precincts. A terra-cotta model of a temple, found at Dali, has in its upper story a multitude of pigeonholes. A dove that was believed to be the messenger of Mohammed used to perch upon his shoulder. And to-day, in the courtyard of the great mosque at Mecca, are more than two thousand of these birds; and to feed them is a duty of all worshippers [March 1898:228].

According to Roman mythology, an eagle was the messenger to Jupiter, the king of the gods. In Polynesia, birds were sent by the gods to warn people of looming danger (March 1898) and along the environs of the Missouri River, local tribes read the signals and messages of birds through the knowledge passed down from generation to generation in storied traditions and personal experiences.

Birds are often heard before they are seen. Their vocalizations are one of their most readily recognizable features and they create a soundscape wherever they are present (Feld 1990). Most birds are social creatures and their calls and responses sometimes seem to imitate human speech. Many bird mannerisms and body language exude an air of intelligence and a keen awareness of their surroundings. Their quick movements and inquisitive nature bespeak a cleverness that many people can relate to as humanlike. In addition, birds are known for their ability to foresee danger or threats. These characteristics not only identify birds as gifted communicators, but also explain their roles as advisors and messengers.

Flight is a skill limited in the natural world exclusively to birds, bats, and certain insects, although there are a few animals that have developed the ability to glide through the air for short distances, such as flying lizards and flying squirrels. Through flight, birds can become familiar with large swaths of land, travel long distances in a short period of time, and survey the happenings below from the proverbial "bird's eye" perspective: "Able to extend their horizon by mounting far up in the air, and having a telescopic vision, their knowledge of the world is proportionately enlarged" (March 1898:209). Birds are also able to observe others without being an obtrusive presence themselves, allowing them to serve as scouts or spies and to relay valuable information to their flock. Their flight, travel prowess, and covert presence make birds an obvious choice as messengers and advisors to people.

Birds are also distinguished by their ability to transcend the boundaries of space that confine humans and other animals. Birds exist on the earth, but can also occupy the sky world, which is home to many important supernatural beings, facilitating communication between these worlds through their flight. Birds convey messages and warnings between the two worlds. Sometimes birds have the ability to speak in human language; other times humans are able to understand bird language; and still other times a bird's presence will communicate a message without sound. These talents place them in a perfect position to offer help to others in need.

Messenger and helper birds in the Missouri River Basin include cranes, swans, geese, gulls, turkey vultures, eagles, hawks, nighthawks, owls, prairie-chickens, meadowlarks, magpies, crows, ravens, sparrows, bluebirds, chickadees, snowbirds, kingfishers, jays, and others that go unspecified in the oral traditions. These birds appear as advisors to culture heroes, tricksters, humans, spirits, gods, and other animals. They are not solely beneficent or maleficent, but take on a variety of roles depending on the bird, the context, and those individuals they are helping.

EAGLES

As noted previously, eagles are particularly well known for their connections to supernatural beings in the sky. As the embodiment of Thunderbird, eagles have unique supernatural powers. To the Crow, eagles carry additional significance as the Sun's messenger to the people, acting as a go-between to bestow the Sun's knowledge and power on individuals who seek it through prayer, fasting, visions, and dreams. The founder of the Crow's Eagle Chapter of the Tobacco Society, Sore-tail, was visited by the Sun, who explained that he would send an eagle as his messenger (Lowie 1919:129–130, 1922b:341). The eagle revealed to Sore-tail a special lodge meant to be used for a Tobacco Dance. He decorated the lodge and the dancers with images of eagles and eagle feathers.

Eagle parts and feathers also play a large role in the Plains Sun Dance because eagles carry messages and prayers to the Sun. The Chief's Pole in the Sun Dance lodge is representative of an eagle's nest (Voget 1995:199). A man with eagle medicine, possibly that of the spotted eagle specifically, would climb up into the nest on top of the lodge and imitate the behavior of an eagle (Lowie 1915b:37). The Chief's Pole was used by the pledger of the ceremony (often someone mourning the death of a loved one lost in battle) and by medicine men with eagle power to send their prayers and receive messages from the Sun (Voget 1995). Both the pledger and the dancers would utilize eagle-bone whistles to carry their prayers to the Sun, who is sometimes also equated with the Creator (Figure 6.1).

The eagle can also be a messenger in its own right, independent from its connections with the Sun or Thunder spirit. In a Crow story, a man gambles away all of his possessions, as well as his wife, through losses at the hoop

Figure 6.1. Crow Sun Dance center pole (2010). Photo by Samrat Clements.

game (Lowie 1918:200–202). He asks many birds for help, but only the bald eagle has the power to defeat the father of the man's gambling rival. The eagle instructs the man on how to construct his hoop and what to wager. With the eagle's help, he wins back everything he lost. Birds also act as messengers later in the story, warning the man that his rival was going to attack him and instructing him to make a shield decorated with birds to protect him.

The eagle's place in Crow culture provides a background for understanding why eagles, eagle feathers, and eagle parts are so significant in Missouri River native life and religion. As the intermediaries between humans and gods, these birds help people to improve communication and to ease their hardships or attain a higher level of spiritual power and understanding. It is also important to note that in the upper Missouri the Golden Eagle plays a more significant role than the Bald Eagle in cultural beliefs and practices.

Today, eagle sightings are a sign of good luck and good will. Eagles perched upon or flying over research sites while conducting ethnographic interviews, for example, have helped the authors improve rapport with Native Americans who see the presence of an eagle as an omen of positive things to come out of the interaction.

MAGPIES

'Well stand just where you are and move that long, shining black tail of yours. Move it up and down, and sideways. Twist it in every direction that you can.' The magpie did as he was told, and Crow Man got down on hands and knees, and went around, watching the shifting, wiggling, fanning tail. Suddenly he cried out: 'There! Hold your tail motionless in just that position!' and he moved up nearer and looked more closely at it. The sun was shining brightly upon it, and the glistening black feathers reflected the tree-tops, and the sky beyond them. Long, long, Crow Man stared at the tail, the [Blackfoot] looking on and holding their breath, and at last he said to Lame Bull, 'I can see your daughter, but she is beyond my reach: I cannot fly there. She is up in the sky land and Thunder Man has her!' [Schultz 1916:27–88].

One of the most frequent messenger and helper characters in oral traditions for the region is the magpie (Figure 6.2). Among Plains tribes as far south as the Cheyenne in Wyoming, the magpie is understood as a "sacred messenger to the high god because it comes near to human habitation and overhears their conversations" (Moore 1986:181). Magpies are abundant and gregarious along the Missouri River, and they appear in the stories of many tribes, especially in conjunction with buffalo.

The magpie appears in the Blackfoot origin story of the Bulls Society (Grinnell 1913:68–76, Grinnell 1920). During a time of scarcity, a woman fleetingly promised to marry a buffalo if they would jump off the buffalo pound, or *piskun*, that her village had set up. One buffalo heard her promise and took her away with him, and the woman's father set out to find her with the help of a magpie. Flying over the buffalo herd, the magpie spotted the man's daughter and relayed to her that her father was waiting for her by the buffalo wallow, but the buffalo husband sensed the father's presence and trampled him. The magpie was then sent out by the women to find any remains of the father. The daughter sung over them to bring her father back to life. Impressed by her strong medicine, the buffalo husband

Figure 6.2. Magpie listening to an interview (2012). Photo by Samrat Clements.

allowed her and her father to return to their village, and gave them the songs and dances to begin the Bull Society.

Mandan and Hidatsa origin stories include accounts of a major flood, which was caused by animals conspiring against humans. A boy transformed himself into a magpie to spy on the animals, undetected, and report back to his people (Beckwith 1937:19–20; Bowers 1992:300). The magpie-boy warned the people of the dangerous flood that was coming, saying "The hard time is coming now!" the first time, but he was not heeded. Then he warned them again (Beckwith 1937:19). The magpie-boy guided the men, who had taken the form of buffalo, and his mother Yellow Woman to Bird-Beak Hill, but because they did not heed the magpie's first warning, only Yellow Woman and one buffalo made it to safety in the flood. When warnings and messages from birds are not heeded, those who have not listened almost unfailingly face grave consequences.

The magpie also plays messenger in a Hidatsa buffalo story that explains the origins of the Imitating Buffalo ceremony (Bowers 1992:439–440). An old couple married their daughter to a skilled hunter, who became selfish when game turned scarce in the winter months. The hunter forbade his wife from offering any food to her parents, and they began to starve. Seeing what was happening, the magpies reported this information to a valley where the buffalo where grazing. The buffalo council sent one buffalo to always be with the Hidatsa. This buffalo left a blood clot, which turned into Blood Man, who provided the elderly couple and the village with abundant buffalo meat.

MEADOWLARKS

Meadowlarks often carry messages of warning to others, but also impart valuable advice to people who are in grave danger. Sometimes messages maybe be sent from one person to another through the meadowlark, or the meadowlark may carry warnings of its own volition. Coyote sends a message through a meadowlark to Hidatsa culture hero Charred Body warning of an impending attack on him from a nearby village, allowing him time to gather a defense (Beckwith 1937:28; Parks 1991:314–319). In an Arikara variation on this story (Parks et al. 1978:85), a meadowlark repeatedly warns Speckled Arrow to build a palisade before an impending attack.[1] Speckled Arrow insists on making many arrows instead. These arrows are scattered by a weasel, and when Speckled Arrow returns to his village after chasing the weasel, it had been burned and all of his people slaughtered by three adversaries: Fiery Moccasin, Pot-tilting Woman,[2] and Swallower. Wilson (1908:1) notes that the Hidatsa believe that Burnt Arrow was at times a man, and at times a bird with a white tail and a hawk's body known in the Hidatsa language as *Ic-pa-ta-ki*. He came down from the sky in the form of an arrow with three marks on its sides, which represent lightning, commonly associated with the Thunderbird.

Meadowlarks are sometimes oracles who speak of future events, like in the widespread story of a boy who is carried into the nest of a family of Thunderbirds in order to help them kill a water monster. In Ella P. Water's Arikara adaptation of the story, the boy—Tied Face[3]—was visited by a meadowlark who warned him, early in the story, "This is what is going to happen" (Parks 1991:952). At other times the meadowlark can offer key information to someone in a dangerous situation, as is the case later in the Thunderbird story when, in his battle against the water monster, a meadowlark told Tied Face where to aim in order to kill the serpent (Parks et al. 1978:55–57). After

the serpent was killed, he invited all the birds to come and feed on the water monster's carcass.

In the Arikara story of "Forked Feather," a woman is warned by Four Grandmother Mice that her future husband is going to kill her (Parks et al. 1978:100–101). Just as the woman was about to be pushed by her husband into the river to be eaten by fish, a meadowlark flew by and told her to steal her husband's necklace. The necklace contained her husband's heart, and she kept it with her to subdue her husband. Similarly, in Mandan stories explaining the origin of the Nuptadi and Awatixa Shell Robe Bundles, a meadowlark saved a young man from the deadly clutches of his evil sister (Bowers 2004:369–372). The meadowlark first warned that his sister is plotting to kill him and use his scalp to adorn her robe. Later the meadowlark explained that the only way to kill his sister is by shooting arrows at the magpie feathers on her head; the boy split the magpie feathers (his sister's heart) with his arrow and the sister died, passing on her robe to him for victory in battles.

There are many variations on another well-known story about a woman and her son living in the sky. The boy is warned never to shoot meadowlarks by his father, who is a Star (Bowers 1992:334; McCleary 1997:38).[4] But when a meadowlark begins bothering the boy, he shoots at it. He is told by the bird that he is actually from the earth below and does not belong in the sky world, so he and his mother return to earth through a hole in the sky. It soon becomes apparent that the boy has strong powers and he goes on to complete many holy tasks.

RAVENS AND CROWS

The Blackfoot believe that ravens circling overheard is a signal that a messenger will soon arrive with news from far away (Grinnell 1920:261; McClintock 1999:483). In the Crow story of "Lost Boy" told by Bear's Arm, a woman lost her only son and her husband went out looking for him. After waiting in camp for four days after the rest of the group had moved on, the husband was finally visited by a raven. The raven spoke to him and told him exactly where to find his son and how to help him escape from his captor, along with captives from many other tribes. After the captives were led out, the raven would caw and separate the captives into their tribes with their respective languages.

---

[1] Speckled Arrow is the Arikara equivalent of the character Charred Body. He is also called Burnt Arrow by the Hidatsa (Wilson 1908).

[2] Pot-tilting Woman (Beckwith 1937:37) appears in several Mandan and Hidatsa oral traditions, like "Spring-Boy and Lodge-Boy" (Beckwith 1937) and a similar account recounted by Matthews (1877) of "Long Tail and Spotted Body."

[3] Tied Face is also known as Packs Antelope, or Carries the Antelope in other accounts of a similar story.

---

[4] Sometimes the meadowlark is replaced with ravens (Lowie 1918:165–169). The father is a star, the moon, or the sun, depending on the tribe and storyteller.

According to Bear's Arm, this raven is the namesake of the Crow Indians. French explorers mistranslated an Absaroka term, calling the tribe People of the Crow rather than People of the Raven (Voget 1995:xvii).

In the Hidatsa story of the Sunset Wolf ceremony, Hungry Wolf was out eagle trapping when he was captured by a group of Cheyenne, who tortured him and left him attached to posts for three days (Bowers 1992:410–411). On the fourth day, a raven observed the tortured man and notified the Chief of the Wolves at the "spot where the sun sets." They turned themselves into humans to save him and gave him the Sunset Wolf ceremony. Raven feathers are worn during performances in the ceremony in honor of the raven. The raven appears in another story of torture, in which a Mandan fasted and tortured himself at the base of an oak tree where a raven lived in order to receive the name of the Arikara man who killed his brother (Bowers 2004:166). He made an offering to the People Above and the ravens nesting in the oak tree, after which the raven revealed to him the name of the person who had killed his brother. In this case, the raven not only foresees the future, but can also reveal information about the past in order to help those with raven medicine.

The very presence of a raven is a message in itself. In a Blackfoot story, a woman promised in her mind to marry one of the buffalo if he would drive the rest of the herd over a cliff to provide food for her village (Wissler and Duvall 1908:113–116). The woman was out collecting wood sometime after this when she noticed two ravens sitting near her. She thought it was strange, but ignored them and continued in her work. Then the ravens began to circle her head, warning her that a buffalo man was coming for her. By then it was too late, and the buffalo man, Scabby-Bull, brought birds—"swallows, small birds, canary birds, etc."—to swirl around her head and confuse her until she agreed to marry him and they went off together. From this story one learns that birds can have conflicting helper roles when bridging communication between different kinds of beings. Birds warn the Blackfoot woman, but also help Scabby-Bull capture the woman. Later in the story, a bluebird, a blackbird, and a prairie-chicken help the woman's husband find Scabby-Bull and his wife, also giving him advice on how to kill the bull with specially prepared arrows and sharing their powers with him as well. This story illustrates the complexities of bird roles in oral tradition.

Not as imposing as ravens, crows nonetheless figure prominently as messengers. Has the Horn, an Arikara woman, and her sister were going to collect buffalo berries on the far side of a river. As they floated across on a bull-boat, a crow swooped down on them so that they would not continue. Has the Horn recognized the warning, but vowed to continue on "no matter what may even happen" (Parks 1991:1233). As the crow warned, the women were attacked by Sioux raiders, but Has the Horn was a powerful woman and chased them away.

The crow also helps an Assiniboine man who suspects that his wife has committed adultery (Lowie 1909:216–217). When the man returns from a hunting expedition to find his wife missing, a crow informs him of her whereabouts and how she could be found. With the help of the crow, he tricks his wife into returning home with him from a neighboring camp. Hereafter the crow acts as a lookout over the man's wife to guard her fidelity. When the wife's lover returns to steal her away, the crow has the husband tie his wife to a treetop, out of her lover's reach. The husband continues to punish his wife for her unfaithfulness until she finally agrees to behave; then the man thanks the crow for his help and dismisses him.

CHICKADEES

Despite its small size, the chickadee is valued by the Crow and the Blackfoot for its wisdom and power. As a wise bird, the chickadee is also valued as a messenger whose instructions should be heeded. Pretty-shield's grandmother was adopted by a female chickadee who told the grandmother to meet her by the creek (Linderman 1972). The chickadee told Pretty-shield's grandmother that although the chickadee was her friend, she would have to hurt her first.

> "I am a woman, as you are. Like you I have to work, and make the best of this life," said the bird. "I am your friend, and yet to help you I must first hurt you. You will have three sons, but will lose two of them. One will live to be a good man. You must never eat eggs, never. Have you listened?" asked the bird, settling down again, and growing small. "Yes, I have listened," my grandmother told that chickadee, and from that day she never ate an egg [Linderman 1972:160].

Pretty-shield's grandmother warned her husband and sons not to eat eggs, but they did not listen. Just as the chickadee portended, two of her sons as well as her husband were killed by Lakota attacks not long afterward. Therefore, Pretty-shield asserted, "It is bad to harm the chickadee, and foolish not to listen to him" (Linderman 1972:160).

Plate 1. Mallard (*Anas platyrhynchos*).
Image from http://www. publicdomainpictures.net/view-image
.php?image=33434&picture= mallard-anas-platyrhynchos.

Plate 2. Common Merganser (*Mergus merganser*).
Image from http:// www.public-domain-image.com/free-images/
fauna-animals/birds/ common-merganser-bird/attachment/
common-merganser-bird.

Plate 3. Canada Goose (*Branta canadensis*).
Image from http://www. publicdomainpictures.net/view-image
.php?image=51043&picture= goose-on-grass.

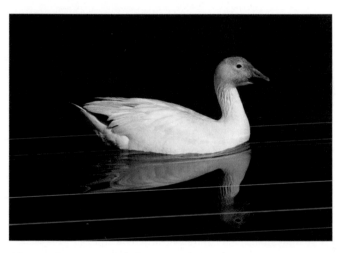

Plate 4. Snow Goose (*Chen caerulescens*).
Image by Cephas—Own work, CC BY-SA 3.0, https://commons
.wikimedia.org/w/index. php?curid=10136226.

Plate 5. Tundra Swan (*Cygnus columbianus*).
Image by Andy Reago & Chrissy McClarren—Tundra Swan,
CC BY 2.0, https://commons. wikimedia.org/w/index.php
?curid=47083771.

Plate 6. Common Loon (*Gavia immer*).
Image from http://www.public-domain-image.com/free-images/
fauna-animals/birds/loon-pictures/ gavia-immer-loons-birds
-pair-swimming/attachment/gavia-immer-loons-birds-pair
-swimming.

Plate 7. Western Grebe (*Aechmorphorus occidentalis*).
Image by Deadcode at the English language Wikipedia, CC BY-SA
3.0, https://commons.wikimedia.org/w/index.php?curid=5828502

Plate 9. American Bittern (*Botaurus lentiginosus*).
Image from Public Domain Images: http://www.public
-domain-image.com/free-images/ fauna-animals/birds/
american-bittern-bird/attachment/american-bittern-bird.

Plate 8. American White Pelican
(*Pelecanus erythrorhynchos*).

Image from http://www.public-domain-image.com/
free-images/fauna-animals/ birds/pelican-birds
-pictures/white-pelican-bird-pelecanus-erythrorhynchos
-high-definition-photo/attachment/white-pelican-bird
-pelecanus-erythrorhynchos-high-definition-photo.

Plate 10. American Coot (*Fulica americana*).
Image from http://www. publicdomainpictures.net/view-image
.php?Image=46735& picture=coot.

Plate 11. Sandhill Crane (*Grus canadensis*).
Image by John J. Mosesso—Public Domain, https://commons.
wikimedia.org/w/index.php?curid=883871.

Plate 12. Whooping Crane (*Grus americana*).
Image Public Domain, https://commons
.wikimedia.org/w/index.php?curid=401689.

Plate 13. Killdeer (*Charadrius vociferous*).
Image from http://www. public-domain-image.com/free-images/
fauna-animals/birds/ sandpiper-birds-pictures/killdeer-animal
-bird-on-coast-charadrius-vociferus/attachment/killdeer-animal
-bird-on-coast-charadrius-vociferus.

Plate 14. Wilson's Snipe (*Gallinago delicata*).
Image by Alan D. Wilson—Nature's Pics online, https://commons
.wikimedia.org/w/index.php?curid=5807412.

Plate 15. Sharp-tailed Grouse (*Tympanunchus phasianellus*).
Image by Gerry from Fort St. John, BC, Canada—Sharp-tailed
Grouse, CC BY 2.0, https://commons.wikimedia.org/w/index.php
?curid=8491686.

Plate 16. Greater Prairie-chicken (*Tympanunchus cupido*).
Image from http://www.public-domain-image.com/free-images/
fauna-animals/ birds/grouse-birds-pictures/greater-prairie-chicken
-tympanuchus-cupido/attachment/greater-prairie-chicken
-tympanuchus-cupido.

Plate 17. Osprey (*Pandion haliaetus*).
Image from http://www. publicdomainpictures.net/view-image
.php?image=87965& picture=osprey-in-the-wild.

Plate 18. Golden Eagle (*Aquila chrysaetos*).
Image from http://www. publicdomainpictures.net/view-image
.php?image=19747&picture= grace-golden-eagle.

Plate 19. Bald Eagle (*Haliaeetus leucocephalus*).
Image from http://www. publicdomainpictures.net/view-image
.php?image=16301&picture= american-bald-eagle.

Plate 20. Cooper's Hawk (*Accipiter cooperii*).
Image from http://www. publicdomainpictures.net/view-image
.php?image=19551&picture= coopers-hawk-128.

Plate 21. Sharp-shinned Hawk (*Accipiter striatus*).
Image by Abdoozy at English Wikipedia, CC BY-SA 3.0, https://
commons. wikimedia.org/w/index.php?curid=7060919.

Plate 22. Red-tailed Hawk (*Buteo jamaicensis*).
Image by Rhys A.—Hawk vs. RabbitUploaded by
Snowmanradio, CC BY 2.0, https://commons
.wikimedia.org/w/index.php?curid=13248275

Plate 23. Rough-legged Hawk (*Buteo lagopus*).
Image by Walter Siegmund (talk)—Own work, CC BY 2.5,
https://commons. wikimedia.org/w/index.php?curid=
3408643.

Plate 24. Northern Harrier (*Circus cyaneus*).
Image by Kositoes - Own work, Public Domain, https://commons.wikimedia. org/w/index.php?curid=7656011.

Plate 25. American Kestrel (*Falco sparverius*).
Image from http://www. public-domain-image.com/free-images/fauna-animals/ birds/hawks-falcons-birds-pictures/american-kestel-bird-falco-sparverius/attachment/american-kestel-bird-falco-sparverius.

Plate 26. Barn Owl (*Tyto alba*).
Image from http://www. publicdomainpictures.net/view-image.php?image=94658&picture= animal.

Plate 27. Burrowing Owl (*Athene cunicularia*).
Image from http://www. publicdomainpictures.net/view-image.php?image=116699& picture= burrowing-owl.

Plate 28. Snowy Owl (*Bubo scandiacus*).
Image by David Syzdek—Snowy Owls (8 of 22), CC BY-SA 2.0, https://commons.wikimedia. org/w/index.php?curid=19899899.

Plate 29. Eastern Screech-owl (*Megascops asio*).
Image from https://www.nps.gov/common/uploads/photogallery/akr/park/cong/36BF641B-1DD8-B71C-07C89D64F4A47B1E/36BF641B-1DD8-B71C-07C89D64F4A47B1E.jpg

Plate 30. Belted Kingfisher (*Ceryle alcyon*).
Image from http://www. public-domain-image.com/free-images/
fauna-animals/ birds/belted-kingfisher-bird-juvenile-female/
attachment/belted-kingfisher-bird-juvenile-female.

Plate 31. American Crow (*Corvus brachyrhynchos*).
Image from http://www. publicdomainpictures.net/view-image
.php?image= 24924& picture=one-black-crow.

Plate 32. Common Raven (*Corvus corax*).
Image from http://www. publicdomainpictures.net/view-image
.php?image=22162& picture=raven.

Plate 33. Clark's Nutcracker (*Nucifraga columbiana*).
Image from http://www.public-domain-image.com/free-images/
fauna-animals/ birds/clarks-nutcracker-bird/attachment/
clarks-nutcracker-bird.

Plate 34. Black-billed Magpie (*Pica hudsonia*).
Image from http:// www.public-domain-image.com/free-images/
fauna-animals/birds/ black-billed-magpie-bird-pica-hudsonia/
attachment/black-billed-magpie-bird-pica-hudsonia.

Plate 35. Mourning Dove (*Zenaida macroura*).
Image from http://www.publicdomainpictures.net/view-image
.php?image= 128899&picture=mourning-doves.

Plate 36. Northern Flicker, red-shafted (*Colaptes auratus*).
Image from http://www.public-domain-image.com/free-images/
fauna-animals/birds/close-view-of-northern-flicker-bird-sitting
-on-end-of-dead-branch/attachment/close-view-of-northern
-flicker-bird-sitting-on-end-of-dead-branch.

Plate 37. Northern Flicker, yellow-shafted
(*Colaptes auratus*).
Image from http://www.public-domain-image.com/
free-images/fauna-animals/birds/northern-flicker
-yellow-shafted-bird-colaptes-auratus/attachment/
northern-flicker-yellow-shafted-bird-colaptes-auratus.

Plate 38. Pileated Woodpecker (*Dryocopus pileatus*).
Image by Lorax at the English language Wikipedia, CC BY-SA 3.0,
https:// commons.wikimedia.org/w/index.php?curid=23541.

Plate 39. Red-headed Woodpecker
(*Melanerpes erythrocephalus*).
Image by Mdf—Own work, CC BY-SA 3.0, https://
commons.wikimedia. org/w/index.php?curid=2075655.

Plate 40. Black-capped Chickadee (*Poecile atricapillus*).
Image from http://www.publicdomainpictures.net/view-image
.php?image= 42749&picture=chickadee.

Plate 41. Mountain Bluebird (*Sialia currucoides*).
Image from http:// www.public-domain-image.com/free-images/
fauna-animals/birds/ mountain-bluebird-sialia-currucoides/
attachment/mountain-bluebird-sialia-currucoides.

Plate 42. Eastern Bluebird (*Sialia sialis*).
Image from http://www. publicdomainpictures.net/view-image
.php?image=49086&picture= birds.

Plate 43. American Robin (*Turdus migratorius*).
Image from http://www. publicdomainpictures.net/view-image
.php?image= 14112&picture= american-robin.

Plate 44. Snow Bunting (*Plectrophenax nivalis*).
Image by Graham Racher from London, UK—IMG_8842,
CC BY-SA 2.0, https://commons.wikimedia.org/w/index
.php?curid=9527129

Plate 45. Yellow-headed Blackbird
(*Xanthocephalus xanthocephalus*).
Image by Dave Menke, in U.S. Fish and Wildlife National
Digital Library, Public Domain, https://commons.wikimedia
.org/w/index.php?curid=486280.

Plate 46. Western Meadowlark (*Sturnella neglecta*).
Image by Kevin Cole https://commons.wikimedia.org/
w/index.php?curid=3892445.

Plate 47. American Goldfinch (*Carduelis tristis*).
Image from http://www. publicdomainpictures.net/view
-image.php?image=108364& picture=goldfinch-feeding.

The chickadee also appears in one of the Hidatsa Grandson (Morning Star) stories, when a man named Yellow Dog wants to steal Grandson's wife (Bowers 1992:337–338). Grandson collects all the animals so that Yellow Dog's village will starve; yet he then receives word that somehow the people have found food. He sends out a chickadee to scout the situation and the bird is trapped, fed, and returned to Grandson to let him know that the people are not starving. In another version of this story, the chickadee is replaced with a mousehawk (a more specific identification of the mousehawk could not be made).

SNOW BUNTINGS

Snow buntings or "snowbirds" in general appear as messengers in two Crow stories recorded by McCleary (1997). "The Seven Brothers" (McCleary 1997) is similar to the Mandan and Hidatsa stories of the Snow Owl myth (Bowers 1992, 2004). In this story, a man named Yellow Leggings was trapping eagles when he became stuck in his eagle-trapping pit. When he got unstuck he found himself in a strange place and was asked to kill a troublesome and dangerous elk by an old man named White Owl. The elk was powerful and guarded by helpers. Yellow Leggings was in a state of despair about how he could possibly kill such a mighty creature when he was visited by a snowbird. The snowbird explained that if he asked the mole people for help, Yellow Leggings would be successful in killing the elk. Following the bird's advice, Yellow Leggings got help from the moles, who brought him underground to shoot the elk from below. He killed the elk and cut off the tip of the its antler to bring back to White Owl, thus completing his first task. The snowbird helped Yellow Leggings throughout the story with the other tasks he is charged with by White Owl, and White Owl eventually shared with Yellow Leggings all the powers of the owl.

In another Crow story, a snowbird helped a virtuous young woman who was going to marry a man called *Iisbishéetbishish*, Worms in His Face (McCleary 1997:84). The man had a history of taking women off with him, but on the third night he would take them to bed and wake up with worms all over his face. The women would run away scared, only to be caught by Worms in his Face and fed to his parents. A snowbird told this virtuous woman that if she rubbed her face against his in the morning, his face would be healed and the man would become handsome again. When *Iisbishéetbishish* awoke on the fourth morning, his face was beautiful and the woman was saved by the advice of the snowbird.

OWLS

Owls are the master messengers, as in the Arikara story when an owl warns a boy that his sister has become a Whirlwind and is planning to kill him (Dorsey 1904:134–136). The owls offer the boy a safe haven in their den so that he can safely negotiate with his sister to save his own life. He offers the Whirlwind his first wife, which appeases the sister. Whirlwind-sister then gives the boy the power of the Whirlwind, which is the power to kill the enemy.

While the owl is sometimes a helpful messenger, the birds can also be manipulated by tricksters for their own purposes, even when it means violating strict cultural taboos. The Crow Indians had a stringent taboo between mothers- and sons-in-law, discouraging any contact or conversation between the two (Lowie 1918:49–50). However, Old Man Coyote tricked his mother-in-law into joining him in the woods, under the auspices of joining a war party, because he wanted to sleep with her (Lowie 1918; Voget 1995). He waited until nightfall, and then he summoned an owl to scare his mother-in-law by making loud noises around their shelter and saying "Sleep together, if you do not do so, you shall die" (Lowie 1918:50). The woman was very frightened and suggested that they do what the owl had told them; Old Man Coyote left the shelter and smoked a pipe with the owl in gratitude before returning and having his way with his mother-in-law. This story is thick with the multilayered roles that birds take on in Indian life. In Crow traditions, the owl can be at once a fearful creature and a healer, a messenger, and the tool of a trickster to get around a strict taboo.

Owls can also carry lucky information. In a story recorded by Wissler (1912:81),

> One time at a place where Badger Creek runs into Two Medicine River, I saw two owls on a tree. Each owl in turn sang a song. Then one of them spoke to me, telling me that I would always be fortunate and get much property. They told me to take some of their children for medicine. So ever since that I have kept the head of an owl and I have always had much property.

WATERBIRDS

Waterfowl, and especially geese, are considered by the Mandan, Hidatsa, and Arikara to be messengers and representatives for Old-Woman-Who-Never-Dies, sent by her to carry the corn spirits and other important plant spirits to the people (Beckwith 1937; Bowers 1992,2004; Peters 1995; Will and Hyde 1964). Ceremonies and offerings were made to waterbirds and their behaviors were imitated in many aspects of the Goose Society rituals and Old-Woman-Who-Never-Dies rites. Geese and other migratory birds connect people and supernatural beings with the landscape and regulate the rhythms of people's lives. In this case, not only do they migrate southward to the place where Old-Woman lives in her large earth-lodge on an island near the mouth of the Mississippi River (Bowers 2004:203), but the timing of their departure and arrival also coincides with the planting and harvest schedules for corn and other important agricultural products.

In Spring, migratory waterbirds like ducks, geese, cranes, and herons arrived, carrying with them the spirits of the corn and other garden products. Their arrival was taken as permission from Old-Woman to begin the planting season (Bowers 2004:245; Peters 1995). Each bird is associated with a cultivated plant, the most important being the geese who carry the corn spirits, the duck (bean spirits) and the swan (gourd spirits) (Beckwith 1937:53). In preparation for the arrival of geese bearing the corn spirits and Old-Woman's blessings, offerings of meat were made to the birds on drying platforms near the villages. The meat was tied to sticks pointing toward the north and south, relating to the directionality of the flight path for the sacred birds (Bowers 2004:340). Acting as couriers between Old-Woman and the villages, the meat offerings would be carried back to Old-Woman-Who-Never-Dies by the birds (Will and Hyde 1964:271).[1]

Waterbirds associated with Old-Woman-Who-Never-Dies are believed to be women who are reincarnated into birds. Today, women of the Goose Society refer to waterfowl as their "sisters," demonstrating a feeling of kinship. When asked if she did anything to prepare for the arrival of Old-Woman-Who-Never-Dies' birds in the spring time, a Mandan-Hidatsa consultant who is a member of the Goose Society explained: "I have my pipe I bought and I smoke my pipe. Grandma-Never-Dies is in the sweetgrass, she's in the wind. So wherever I am, she will feel me and she will smell me."

---

[1] The practice of placing dried meat offerings, as all other Native American religious rituals, was outlawed by the federal government at the turn of the twentieth century (Will and Hyde 1964:271). These practices may have continued in secret, however, it was not until the passage of the American Indian Religious Freedom Act in 1978 that tribes were allowed to revitalize their traditional religion.

OTHER BIRDS

The Cooper's Hawk, commonly known as the chicken hawk, acts as a messenger to those who have chicken hawk medicine. In the tradition of the Mandan-Hidatsa-Arikara, people who have chicken hawk medicine can understand and talk to the bird. One consultant explained:

The chicken hawk, a lot of people would use them when they used their medicine, they would send him out to watch another person, and then [the chicken hawk] would come back and tell. That's what the chicken hawk does. He would tell everything. I have a cousin, she really believed in the chicken hawk because her grandpa had the medicine for the chicken hawk. He would send him out and he would come back and tell, and watch other people for him. So she really believed in it.

—Mandan-Hidatsa-Arikara consultant

The role of prairie-chickens as messengers is an interesting one, considering that these birds are not known for their flying skills. They most often use a short, abrupt burst of energy to take momentary flight. Although the prairie-chicken is often better known for its booming ritual and mating dance, it takes on a particularly honorable role in the Arikara creation story as a messenger for Mother Corn to the gods at each of the four directions and Nesaru, the creator (Dorsey 1904:20). Prairie-Chicken carries the pipe to the gods in the southeast, southwest, northwest, and northeast for Mother Corn, before bringing it at last into the heavens for Nesaru to smoke. Prairie-Chicken persevered through storms and strong winds to carry the pipes for Mother Corn, and the white spots seen on the feathers of prairie-chickens today are said to have resulted from the rocks that hit Prairie-Chicken as he flew through the storms to deliver the pipe.

Although the Turkey Vulture (*Cathartes aura*) is rarely mentioned in regional ethnography. It is regarded by the Blackfoot as a harbinger of misfortune. Birds that subsist on carrion are often associated with death, disease, and bad luck, but their appearance is not always negative. Carrion birds, such as the turkey vulture, may also appear as a warning to preempt trouble. Chewing Black Bones explained to Claude Schaffer in an interview that his father had spotted a turkey vulture while out on the warpath (Hungry-Wolf 2006:136). The leader of the group took this as a sign of misfortune ahead, inferring that the bird was heading them off and telling them to return home. Most of the group returned home safely, but of the six that continued on the warpath, only one returned alive.

Several other birds are mentioned in stories as helpers to a character in need. In these cases, it is not clear that the particular bird in any given story holds special significance as a messenger beyond its being a bird. In a Crow story, for example, a sparrow helps a boy kill Red-Woman, a powerful woman who is pursuing the boy and a young woman (Lowie 1918:125–126). Red-woman had already killed the young woman's seven brothers, but the sparrow explains to the boy that the only way to kill Red-woman is to shoot at the owl feathers she had tied to her head rather than shooting at her body. Red-woman's heart and lungs were located in the feather. Shooting her there, the boy killed Red-woman and was able to bring the seven brothers back to life using a sweatlodge ceremony.

In a Blackfoot story, the kingfisher is the only bird that is able to tell Old Man Napi where he can find his companion Wolf Chief, who had gone missing (Grinnell 1920:151). The kingfisher not only informed Napi that Wolf Chief had been killed by Chief Bear, but further reported to him where and when the bears could be found. According to a Blackfoot consultant, Napi's revenge on the bears caused the second flood, after which Napi recreated the earth. In a Mandan origin story, three swans help direct Lone Man on his journey northward to Dog Den Butte, where Hoita, a speckled eagle, had trapped all the earth's animals (Bowers 2004). In an Assiniboine story, a gull helps an abandoned boy to return home by instructing him to kill a gull and use its skin and feathers to fly home (Lowie 1909:150–151). The gull also helps the boy to pass safely through two treacherous chasms by throwing fish into the cracks, and he returns to his mother unharmed.

BIRDS, GENERALLY

There are also cases where a bird may only be identified in the most general way. In some instances, birds are described in a nonchalant and broad manner, as in the story of "When the Bears Attacked the Arikaras." Alfred Morsette narrates, "Then this young man, as he was going around, was warned by a bird, whatever kind it was" (Parks 1996:179). The kind of bird, in this case, is subordinate to the role of birds on the whole as messengers. In many other stories, messenger birds are simply referred to as "little birds" that come to a person with news or instructions (Lowie 1909:150–151, 1915b:13–14, 1918:152–156; Grinnell 1920:153–154; Wissler and Duvall 1908:71).

Although birds are not always identified by name, this does not necessarily take away from the importance of their role as messengers in oral traditions.

CHAPTER SEVEN

Bird Imagery in Material Culture

Imagery in art and material culture is an evocative way to represent central mythological and spiritual figures, as well as important historical events, people, resources, and natural forces. It is also a means to bring the power or essence of a bird into action by virtue of inscription or representation. When written language is not present, images used on their own or in conjunction with oral communication can serve as one of the most effective ways to convey information to others, whether to express identity, share news, or tell a story. Imagery is used in the adornment of everyday objects and clothing, but also takes on a paramount role in the ceremonial and spiritual lives of Native American communities. Bird imagery and symbols appear on the material culture of the Missouri River people from the prehistoric period to the present.

As early as 1694, in a letter to Jesuit Father de Lamberville, his missionary friend Jesuit Father Marest remarked on the charcoal tattoos of the Assiniboine Indians he encountered: "The Assiniboines have large drawings on the body representing serpents, birds, and various other figures which they print by pricking the skin with little, sharp bones and filling up the holes with wet charcoal dust" (Tyrrell 1931:124). Although the meaning behind the tattoos was not explored in the letter, the missionary's observation provides a glimpse of how entrenched birds are in Native life. Tribal elders still get tattoos that depict individual ritual power (e.g., Zedeño 2008a: figure 5). Bird iconography can be found in everyday objects like clothing and moccasins, as well as objects with power and ceremonial significance such as sacred bundles, shields, and regalia.

Images may be used to communicate kinship and societal ties as well as individual and corporate relationships with the supernatural. The images themselves sometimes hold power and agency. An object may have an animate quality, or what Ananda Coomaraswamy (2007:200) calls "mystical participation." An object "is not only what it is visibly, but also what it represents. Natural or artificial objects are not . . . arbitrary 'symbols' of some other higher reality, but actual manifestations of this reality." An object has the ability to both signify and manifest spirituality and spirit beings. Lohmann's (2003:179) definition of animism as "attributing spirits (foundational sentient agency) to materials" can be pushed further to consider the role of imagery as the embodiment of a soul. This in turn lends the object on which the image is depicted its particular ability or power. People acquire the right to depict powerful beings through a vision, dream, or other experience. He or she thus acquires the ability to give an object its animacy by evoking the supernatural in that image. Animate objects are treated as a group distinct from everyday inanimate material items because of their unique ability to influence human action and experience. People interact with animate objects h in a manner similar to the way they interact with another person or supernatural being (Mills and Ferguson 2008; Zedeño 2008a, 2008b). This is not to say, however, that ordinary things lack a soul. Their animating power may be latent and can be awakened through prayer and song or be transformed by painting a powerful image on them (Zedeño 2009).

Imagery created with paint or pigment can hold further significance because of the "fixed intrinsic" quality of paint as an animating power (Zedeño 2008a:374). When a ritual practitioner applies paint to an object, especially red ochre, that object comes to life—a consecratory act. This is true even when a painted object, such as an arrowhead, does not depict an image. When spirit images are painted onto an object, it may take on additional power from the supernatural being it depicts. For example, Assiniboine warriors would carry shields known as *woha'tcanga*, which might be decorated with the image of a spirit guardian, such as the Thunderbird, from which the warrior had received a vision or dream (Lowie 1909:31). Painted with red, black, or blue paint, the image of a supernatural guardian further protected the warrior beyond the normal ability of a common shield because it evoked protection from the spirit manifested by the image. At times bird imagery may be used simply for aesthetic decoration.

The variety of contexts in which bird imagery appears bespeaks the range of roles that birds play in the lives of individuals and communities. Some images are purely naturalistic, while others are created in the likeness of supernatural creatures and other-worldly beings. Some depictions are so general as to preclude association with a particular kind of bird, while other birds are specifically, though not necessarily realistically, represented for the purposes of summoning or depicting spiritual power associated with that bird (Zedeño 2008a: figs. 1, 4–5). Avian imagery often does not directly resemble the exact physical attributes of a bird, making it hard to assign a taxonomic description to an image beyond the most general terms without understanding the cultural context underlying the imagery. Even considering the cultural milieu, a particular species may not be pinpointed.

EVERYDAY MATERIALS

Although clothing and objects of everyday use in the past were, for the most part, plain and undecorated, birds or feathers may occasionally be depicted on these items. More often, however, decoration was reserved for special objects used in ceremonies or received through special circumstances such as a dream or encounter with a spirit animal. Bird imagery sometimes appears on moccasins, bags, cradleboards, ladles, and baskets through representation in quillwork, beadwork, embroidery, sculpture, and paint (Figure 7.1). It is almost impossible to trace the patterns of style and usage of bird imagery in apparel and quill-embroidered items into the past beyond European arrival in the Americas, because the materials are fragile and often do not survive well in the archaeological record. Therefore, descriptions of prehistoric imagery in Native clothing and adornment rely on oral tradition, the memories recorded by early American ethnographers, and early examples of clothing and stylistic motifs that are in museum collections. Early historic examples also reflect bird and feather

Figure 7.1. Top: Assiniboine moccasins with bird motif (date unknown). American Museum of Natural History (AMNH), Division of Anthropology, 50/ 1969 AB.
Center: Crow moccasins with feather motif of dual-tones lozenges (date unknown), National Museum of the American Indian (NMAI), Smithsonian Institution, 10/4458.
Bottom: Mandan/Hidatsa moccasins with feather motif (date unknown), AMNH, Division of Anthropology, 50.1/5398 AB.

motifs that were popular in the past, but have been adapted to fit the present tastes of Native communities.

Sometimes bird and feather motifs are incorporated onto clothing or objects as a central accent piece to the overall design, as mentioned in Lowie's (1909) descriptions of Assiniboine material culture and dress. Although moccasins used in daily activities were unembellished, those used during special occasions such as dances and ceremonies would be more intricately decorated with quillwork or beadwork. In addition to popular geometric, cross, tent, and angle-across patterns, bird designs have been identified as one of the "more common patterns" to appear on moccasins, and stylized feather patterns appeared in the form of dual-toned "lozenges" (Lowie 1909:19–20). At other times, bird imagery is not central to the design but is just one of several elements. A Sioux cradle decorated with quillwork depicts two rows of bird figures in the midst of other animals and geometric patterns (Ewers 1981:261). According to the Sioux tradition, men's objects are decorated with anthropomorphic and zoomorphic bead and quill designs, while geometric patterns are reserved for use on women's items (Wallaert 2006:7).

CEREMONIAL REGALIA AND ADORNMENT

Bird imagery is employed in important ceremonial proceedings and performances along with paint, tattoos, feathers, and other bird parts. During dances and other ceremonies, participants might paint themselves to resemble a certain type of bird, physically imitate the behavior of a bird, or display images of birds or bird parts associated with the participant personally or with the ceremony being performed. Some imagery in regalia has legendary origins. Other uses of imagery may be received through a dream or transferred from a person with bird power. It is important to note that the rules of ownership of an image that depicts power were very strict, as misuse or unauthorized copy would bring misfortune. When Walter McClintock asked his Blackfoot friends to paint his tipi, they were consternated and told him that could only be done when and if he received a specific vision. Eventually, the Blackfoot ceremonially transferred a tipi design to him so that he could have his tipi painted in the right way (McClintock 1999:211). Blackfoot elders note that these rules are enforced today among traditional "hard line" practitioners, although people without proper cultural knowledge may not follow those rules.

The suit worn during the Blackfoot Okan or Sun Dance ceremony has divine origins that trace back to the culture hero Scarface and his journey to the Sun (Wissler and Duvall 1908:61–65). On a quest to get rid of his disfigured face, Scarface travels to the Sun's lodge to ask for help. When he saves Morning Star, the Sun's son, from the attacks of seven "terrible looking" cranes, the Sun gave Scar-Face a bundle containing a painted shirt and leggings and the sweatlodge ceremony. The shirt bears seven black stripes on each sleeve in representation of the seven cranes Scarface killed, and the stripes are occasionally accompanied by cranes' feet painted onto the shirt. The pants are similarly painted with seven bands on the legs to represent the seven evil cranes.

Representations of birds pervade the procession of the bands into the Arikara Medicine Lodge during a ceremony where members of each band are dressed and painted to represent their band's animal identity (Curtis 1909:64–65; Will 1934:17–18). Three of the nine bands represented in the Medicine Lodge ceremony are bird bands: *Kohnít,* an "unidentified swamp bird" (Curtis 1909:64), *Hwatkúsu,* meaning "Big Foot" or Duck, and *Whúdhu* or Owl. The *Kohnít* band is suggested to possibly represent Cormorant, Crane, or Bald Eagle (Will 1934:17).[1] *Kohnít* band members paint their heads and necks white, with a bluish paint applied to their bodies and red markings to the forehead and back, most closely resembling the appearance of a Sandhill Crane. The Duck Band (*Hwatkúsu*) wore necklaces of duckbills (Curtis 1909:67), and the Owl Band (*Whúdhu*) was painted brown and white with spots "to represent the mottled appearance of the owl, with circles about the eye" (Will 1934:18). The other bands, although representing other spirit animals, often used feathers and down in decorating their regalia.

There are other examples of bird imagery in ceremonial attire: The leader of the Arikara White Crane Dance (*Pai-hun-ghe-nah-Wah-che*) wears a robe made of elk-skin featuring a two-headed crane painted on it (Denig 2000:169); the Assiniboine Fool's Dance (*Wintgon'gax Wacipi*) uses masks that are often decorated with birds in flight and rainbows; and the leader of the ceremony

---

[1] This term appears as *Gohnits* in Will (1934:17–18). He believes this band represents a cormorant rather than a crane or bald eagle. Considering the migratory and habitat ranges of various cormorant species in North America, this would most likely be the Double-crested Cormorant (*Phalacrocorax auritus*), which migrates along the Missouri River and breeds in the summer throughout Montana, North and South Dakota, and parts of other states in the Missouri River Basin. However, the painted markings of the Kohnít more closely resemble the Sandhill Crane with its bluish-gray body, red forehead, and red markings on the body. This assumes that the paint is being used to create bird imagery, although paint colors themselves have certain meaning and the body paint may not be meant to visually resemble the band's namesake.

Figure 7.2. Four Horn's copy of Sitting Bull's drawing of himself with shield in battle with Whites, rescuing Jumping Bull, 1870. National Anthropological Archives, MS 1929A (011).

imitates an eagle feeding her young as she throws food to dance participants (Rodnick 1938:52–53). In preparing to transfer the ceremony of the Big Bird Bundle in Hidatsa Big Bird rites, Smells, an elder conducting the ceremony, painted lightning zigzags all over his body and face and wore eagle claw wristlets in representation of the Thunderbird (Bowers 1992:367). In the Mandan Buffalo Dance, two of the dancers represent birds: the Bald Eagle and a species of hawk called *ictataki* (Wilson 1909:7–8). The Bald Eagle dancer was painted black all over except for his chest and head, to which white paint was applied. The hawk dancer was painted yellow with black spots, and "his hair was gathered in a knot over his forehead with a tuft of leaves on it" (Wilson 1909:8).

Beyond the temporary paint and regalia, tattoos are also a way to permanently display the significance of birds to ceremonial and spiritual life through time-honored designs. A current Medicine Pipe Bundle holder of the North Peigan Nation has a tattoo of thunderbird tracks and hail, a "distinctive design . . . which has been faithfully transferred from robe to blanket and to other objects over untold number of years" (Zedeño 2008a:371–372). The thunderbird-hailstone motif appears throughout the upper Missouri Region and Plains and constitutes an example of persistent imagery. Bird imagery continues to change, and new designs and patterns are emerging as a

reflection of new visions as well as contemporary understandings and aesthetics.

WARFARE

Among the Blackfoot, warriors wore their society or bundle bird insignia. Their horses, too, were painted according to the ceremonial office of their owners (Scriver 1990). Shields used in battle generally display bird imagery, partially due to the many spirit birds that possess skills valued in battle, bravery, and success on the warpath. Shields are a medicine object and animals and birds are often depicted on shields or incorporated through animal or bird parts such as claws or feathers. Sitting Bull, a famous Hunkpapa Lakota spiritual leader who foresaw the defeat of Custer's Seventh Cavalry at the Battle of Little Bighorn, was known to carry a shield painted with the image of a bird and adorned with feathers. In a collection of ledger drawings by Sitting Bull now kept in the National Anthropological Archives, Sitting Bull illustrates himself with this distinctive bird shield in almost every image of battle or raiding (NAA 1929). The shield depicts the image of an eagle or Thunderbird in a blue or green background. The buffalo attached to him through life lines also adds to the power of the warrior (Figure 7.2). Sitting Bull's shield, however, is by no means unique.

Figure 7.3. Thunderbird buffalo hide shield cover from the Crow Indian Reservation, c. 1907. The Thunderbird figure has been painted on and then embellished with feathers. AMNH, Division of Anthropology, 50/6832.

Figure 7.4. War shield associated with Crow leader Sore-Belly. The bird head on the shield is identified as a Wood Stork (*Mycteria americana*). Feathers are also attached to the shield (date unknown). National Museum of the American Indian (NMAI), Smithsonian Institution 11/7680.

Assiniboine war shields, called *woha'tcanga*, were often painted with an image of a supernatural being with whom the person had a connection, such as Sun, Moon, Buffalo, or Thunderbird (Lowie 1909:31; Rodnick 1938:42). Similar themes are present in other tribal shields, such as one described by Crow warrior Two Leggings. It incorporated many different elements of the Thunderbird by showing "pictures of the sun, rain, clouds, and an eagle with lightning striking from its claws" (Nabokov 1970:25).[1] The Crow believe in another world just like the earth called "the Other Side Camp," in which important supernatural beings such as Sun and Moon live with their bird helpers. In this case, the dreamer may have only received a vision of the eagle, but also included on the shield Sun, Lightning, Wind, and Rain because they are a part of the eagle's Other Side Camp clan. The image is a connection not only to eagle medicine but also to the Other Side Camp, an important part of Crow cosmology (Figure 7.3).

Protective images on shields also appear in traditional narratives of the Crow, such as the story of "The Gambler Befriended by Birds," where an unlucky gambler receives help from an eagle and other birds in order to defeat his adversary at playing the hoop game (Lowie 1918:200–202). When the man's rival becomes suspicious of his luck and plans a retaliatory attack, birds appear to the gambler in a vision and instruct him to make a shield decorated with birds. The bird shield protects the gambler from his rival and his rival's powerful father. Bird helpers may also be used in shields if their power helps the shield bearer in battle (Figure 7.4).

Bird imagery is also common on other materials of or related to warfare, such as war shirts and commemorative robes. In addition to the war shield, war shirts were worn by some of the more renowned warriors and were similarly decorated with designs and images received in a dream or transferred from one person with war power to another (Rodnick 1938:42).

Painted images of feathers are another power-laden icon used in war-related attire. Feathers can hold a connection to birds both generally and to specific species. But they may also lend powerful qualities in and of themselves apart from their association with a particular bird, as is the case of the Blackfoot Thunder shield, which depicts the group's characteristic Thunderbird image surrounded by feathers (Figure 7.5). Painted feathers are a symbol of war honors, given to warriors for counting coup or other acts of valor. The "feather circle" design is a popular motif for war honors, meant to represent both the sun and the eagle

---

[1] "Lightning strikes" coming from human or animal bodies do not necessarily depict actual lightning phenomena but, rather, symbolize power emanating from a body.

Figure 7.5. Wolf Collar's shield depicts a thunderbird in the Blackfoot design (date unknown). Sketched by M. N. Zedeño from *Artscanada* 1977:39.

Figure 7.6. Blackfoot handheld drum painted with the image of a black crow (date unknown). Source: Scriver (1990:96).

feather headdress worn by the most successful warriors; the feathers are represented geometrically through elongated triangles arranged and painted like feathers (Gilman and Schneider 1987:51; Schneider 2004:125). Bird feathers are often sewn onto shields, furthering the connection between birds, bird imagery, and warfare.

DRUMS AND OTHER CEREMONIAL OBJECTS

Drums

Brings-down-the-Sun related a story about his father, Running Wolf, to McClintock (1999:425–426):

> He heard the beating of a drum, and, after the fourth beating, was able to sit up and look around. He saw the Thunder Chief, in the form of a huge bird, with his wife and many children around him. All of the children had drums, painted with the green talons of the Thunder-bird and the Thunder-bird beaks, from which issued zig-zag streaks of yellow lightning. . . . Finally the Thunder-bird spoke to my father, saying, 'I am the Thunder Maker and my name is Many Drums (expressive of the sound of rolling thunder). You have witnessed my great power and can now go in safety.

Drums are an important part of many ceremonies as the mouthpiece of communication with the supernatural world, used by humans and spirits. Drumming is strongly tied to the Thunderbird. According to some Plains tribes, thunder is created by Thunderbirds drumming after a flash of lightning (Lowie 1924:340; McClintock 1999:425–426; Parks 1991:400–405). Thunderbird and his incarnation are variously depicted on drums. For example, an eagle appears on one of the drums for the Horse Society rituals, one of the most important ceremonies of the Assiniboine tribe, and is used in conjunction with an eagle bone whistle to ask for health and good fortune for the whole band (Rodnick 1938:50–51). An eagle feather is painted on a large drum used by the Deer Band of the Arikara in the Medicine Lodge ceremony.

Imagery on drums is also associated with other important birds (Figure 7.6). In the Blackfoot Medicine Pipe ceremony, drums are decorated with bird figures or other animal and celestial figures, and the sound of the drum is meant to evoke the drumming of the grouse that had given his power to the Medicine Pipe. The grouse drums its wings against fallen logs in order to attract females to mate, and the sound may be heard over quite a long distance.[1] Goose tracks are also painted onto the drum used in the Old-Woman-Who-Never-Dies ceremonies to represent the goose as one of the central messengers of Old-Woman-Who-Never-Dies (Bowers 1992:345; Peters 1995:112).

---

[1] By using the pile-sort method, Hungry Wolf identified five species of grouse: the Ruffed Grouse (*Bonasa umbellus*), the Spruce (Franklin) grouse (*Falcipennis canadensis*), the Dusky (Richardson's) grouse (*Dendragapus obscurus*), the Sharp-tailed grouse (*Tympanunchus phasianellus*; Plate 15), and the Greater Sage Hen or Sage Grouse (*Centrocerus urophasianus*).

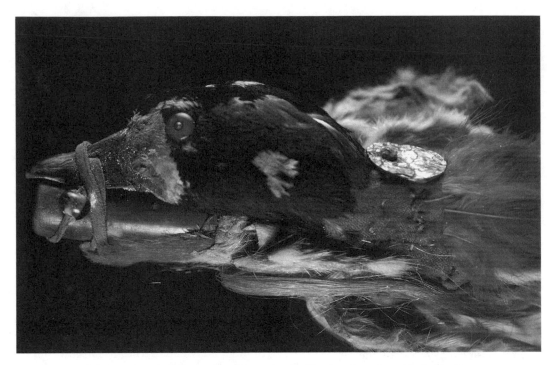

Figure 7.7. Harlequin Duck (*Histronicus histronicus*) affixed to Last Star Medicine Pipe. Source: Scriver (1990:281).

Turtle drums are one of the most significant objects to the Mandan, appearing in the Buffalo Dance portion of the Okipa Ceremony. These were conceived in a dream that instructed a man that there should be four turtle drums rather than one (Bowers 2004:161). In his dream, the four turtle drums were adorned with eagle feathers. Lone Man advised the dreamer that it would take a long time to acquire enough feathers to decorate the drums. Lone Man returned in four years when there were enough feathers and helped the man decorate the turtle drums for the Buffalo dance. They gave the first three drums feathers from the speckled eagle; however, they thought that the last turtle would like calumet feathers best.

The people thought, "We must make some feather offerings on these turtle drums." So they made offerings of big bunches of spotted old eagle feathers, which they tied on the head of each turtle so that they spread forward over the head. For the biggest turtle drum, of whom the Mandan said, "He is chief," they saved their best feathers—the black and white feathers of a young war eagle. The chief turtle then said to the people, "You people have not done right. You have offered the other three turtles old and good eagle feathers, and you offer

but snow-birds' feathers and not eagle feathers" [Wilson 1909:14].[1]

The last turtle was offended, even though the man explained that calumet feathers were "the best of all" (Bowers 2004:161). This turtle returned to the waters of the Missouri near where Fort Yates is now located in North Dakota. To this day, the turtle drums are held within the community in Fort Berthold. At least one turtle drum is still in the possession of the Benson family (Mandan) from Twin Buttes.

Wooden Pipes

Wooden pipes are often carved to represent animals of central importance to a tribe because of their association with the supernatural, with origins of the tribe, or with origins of a particular band or ceremony. Pipes are an integral item to every ceremony as well as personal prayer. Carved bird-head pipes are found in George

---

[1] Wilson (1909:14) explains: "The tail feathers of a war eagle of the second year are pure white with deep black tips. Each succeeding year the white portion fades to a darker mottled color. Snow bird is spoken of contemptuously by the turtle."

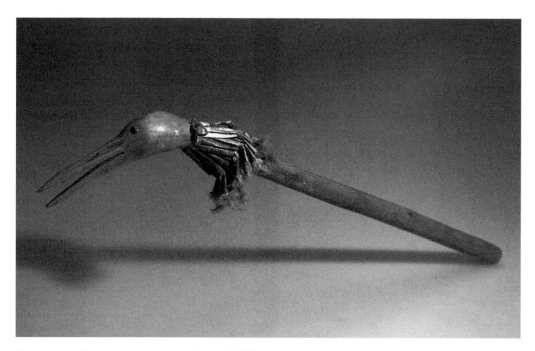

Figure 7.8. Crow dance stick, c. 1907. AMNH, Division of Anthropology, 50/6827.

Catlin's paintings of northern Plains tribesmen and their paraphernalia. They often appear adorned with feathers, cementing the connection of both birds and pipe smoke as messenger beings to the spirits. Denig's sketches of Assiniboine pipes show a bird head, identified as either a Mallard duck or a Red-headed Woodpecker, attached to the end of a calumet pipe that is adorned with three sets of tail feathers from the "war eagle," the Golden Eagle (Denig 2000:Plate 68). Generally, bird heads, particularly waterfowl heads, were fitted to pipe stems to make a calumet or dancing pipe Blackfoot ceremonial pipes also had bird bodies affixed to the stem (Scriver 1990) (Figure 7.7).

Two bundles used in the Corn ceremonies—the Old-Woman-Who-Never-Dies Bundle used by the Hidatsa and the Good-Furred-Robe Bundle used by the Mandan—contain pipes carved into the likenesses of goose heads and bills. These images represent the role of the goose as signaling the beginning and end of the growing season (Bowers 1992:344–346). The wooden pipe of the Robe Bundle is said to have been used by Good-Furred-Robe himself. It is intricately carved with the stem representing a goose's head and neck; the mouthpiece is carved as a goose's bill (Bowers 2004:184–185). These bundles include blackbirds, considered gardening friends because they eat the insects in the fields. The Robe Bundle also contains the head of a green duck. Each of the objects included in these

bundles relate to the storied events surrounding the gift of corn to each of the earth-lodge tribes. Waterbirds and blackbirds play a large role in these ceremonies because of their strong connection to agriculture.

Ceremonial Staffs

More bird imagery appears in carvings on ceremonial staffs. For example, the Crow Hot Dance (transferred from the Hidatsa) included the use of a long wooden staff with a crane's head carved into its end, adorned with bead embellishments gathered around the neck (Figure 7.8; Lowie 1954:131). Another bird carving appears on the flat staffs that are a part of the bundles used in the Okipa ceremony of the Mandan. Each staff displays the thunderbird and the moon carved on one side, and the sun and a star on the other. Bird staffs are a key component in many oral traditions of Missouri River tribes, carried by culture heroes as a symbol of holiness, as a source of power or protection, or as proof of success in war.

Medicine Rocks

Medicine rocks, called *bacóritsi'tse* by the Crow, are part of the Crow tradition and are also important for other plains tribes. *Bacóritsi'tse* are often shaped like figures that represent significant elements of the Crow world such as birds, buffalo, or horses. One of these *bacóritsi'tse*, regarded as

Figure 7.9. Crow Lodge, Blood Reserve, Alberta, c. 1894. Source: Glenbow Archives NA-668-16.

especially remarkable for the dual imagery on either side of the rock, is shaped like a buffalo with a bird on its back on one side, and a little person riding a horse on the other. Bird-shaped *bacóritsi'tse* hold great importance for the Crow and are inherited, rather than sold, through family lines (Lowie 1922b:388).

Painted Tipis

A white man looking upon the inside circle of Painted Tipis, in the great encampment of the Sun-dance festival, would be impressed with their imposing array and with the spectacular effect of their novel colourings and fantastic decorations. But, it probably would never occur to him that he was looking upon pictorial representations of the tipi-owner's religion. As the wearing of the crucifix is the outward sign to the world of the inward faith of many Christians, so these tipi representations of the Buffalo, Beaver, Elk, Otter, Eagle, and Antelope proclaim the belief of the Blackfeet, that these sacred animals and birds have been endowed with power from the Sun, and, therefore, that the owner and his family may secure from them aid in danger and protection from sickness and misfortune. Just as patron saints are worshipped to-day, and the Lares and Penates of pagan Rome were worshipped two thousand years ago for household protection, in like manner the spirit of the otter, buffalo, or beaver, is worshipped and its visible representation on the tipi is held sacred by the Blackfoot family as their powerful protector [McClintock 1999:223–224].

Painted tipi covers may be the most visually striking display of imagery for many Plains tribes, who sometimes painted images of important events or figures on buffalo hides or canvas that formed the outer cover of the tipi. Images on tipis often commemorate courageous acts of war, depict hunting scenes, or show important spirit animals such as the buffalo or the Thunderbird. Among the Blackfoot, tipi designs were originally given to the people by the water animals (McClintock 1999). When a person receives power from an animal through a vision or dream, they also receive instructions for a song, dance, or ceremony. In that vision or dream, they may also receive a design for a special article of clothing, a robe, or a tipi.[1] Ravens, eagles, crows, and other raptors are most commonly pictured on tipis, perhaps because of their associations with war and their status as high power birds.

Decorated tipis appear most frequently in the historical paintings and photographs of the Blackfoot Confederacy divisions, and birds are a regular sight painted onto the tipi-covers of both ceremonial and personal lodges. Grinnell's (1913:198–199) descriptions of painted Blackfoot tipis suggest that the raven "procession" was a common design painted onto the lower border of a tipi, although other bird and animal figures were also popular depending on one's medicine and experiences (Figure 7.9). The Thunderbird spirit also appears on some painted tipis, taking on an otherworldly form that less resembles any earthly bird than a unique spirit being (Figure 7.10). The Thunderbird is often pictured with lightning bolts or hailstones, or may be represented by four claws or four horizontal lines that symbolize Thunderbird tracks or "thunder-trails" (McClintock 1999:218). Importantly, as McClintock notes, only about 10 percent of the people owned a painted tipi; in general, chiefs and priests were the only ones with rights to own one.

According to a Crow consultant, the tribe traces the origin of tipis back to a gift from "a white owl," likely a Snowy Owl or possibly a Barn Owl, giving the tipi itself a sacred connection to the animal or bird world and religious significance (Zedeño et al. 2006:225). The consultant also stressed the point that "we each have our own tipi and story that goes with it." Although Lowie (1922b:401) suggested that painted tipis were not as common among the Crow (he remembered "but a single painted tipi, which was decorated with the figure of the Thunderbird"), the origin story and personal stories connected with tipis frequently draw close connections to the bird world and bird spirits.

Figure 7.10. Blackfoot Thunder Tipi, Browning, Montana, c. 1910. Source: Glenbow Archives NA-3324-1.

In the story of "The Crow Who Went to the Birds' Country," a young Crow man was taken into the sky by the Crane chief and arrived at the sky village of the birds. Each bird had a tipi decorated according to their own colors and markings. The meadowlark's tipi was painted yellow like its breast with a black top. The bluebird's tipi was painted entirely blue. The nighthawk's tipi displayed an image of that bird. And the yellow-crane's tent was large and completely yellow (Lowie 1918:159–160). The incorporation of painted tipis in oral traditions suggests that the practice of painting tipis was more widespread than Lowie observed in his visits to the Crow.

The Thunderbird tipi that Lowie recalled as the only painted tipi may have been associated with the story of Burnt-face (Lowie 1918:152–156), which is similar in many ways to the widespread story of a culture hero who helped Thunderbird defeat a dangerous water monster. Unique to the Crow story, however, the eagle gave Burnt-face eagle

---

[1] It is important to point out that not all tribes along the Missouri River lived in tipis or painted them. The Mandan, Hidatsa, and Arikara built earth lodges that did not lend themselves to painted decoration in the same way that the tipis of the Assiniboine, Blackfoot, and Crow did.

Figure 7.11. Blood Sun Dance lodge frame, 2005. Photograph by M. N. Zedeño.

medicine with the instructions to build a large tipi with the image of an eagle painted onto it. This tipi gave Burnt-face the power to foretell the weather and storms. The centrality of the eagle and the Thunderbird spirit in Crow culture may have also made it a more likely candidate than other birds or animals for display on tipis.

Another example of a tipi design given from a supernatural source, *Es-to-nea-pesta* or the Maker of Storms and Blizzards in the Blackfoot tradition, the Snow Tipi has been passed down through the generations and is known for its power (McClintock 1999:133–138). The Snow Tipi is decorated with four green claws and yellow legs, which represent the Thunderbird, with the top painted yellow like sunlight and a red disc on the back symbolizing the sun itself. The Snow Tipi is so strong that "whenever it is pitched, cold weather and winds are sure to come because of its great power" (McClintock 1999:133). It was acquired by Sacred Otter when he and his son became trapped in a blizzard while running buffalo on the plain. Cold Maker showed him the tipi design and special medicine, which

was transferred repeatedly for protection during bad winter weather. Medicine men also pitched the Snow Tipi when battles were not going the Blackfoot way and their warriors needed a weather break to regroup.

The Sun Dance Lodge

Entire books could be devoted to discussing and analyzing the symbolism and imagery of eagles in the Sun Dance ceremonies of the Plains tribes. Eagles are incorporated into almost every aspect of the ceremony, including the structure of the lodge, which was made to resemble an eagle's nest; carved and painted eagle images; eagle feathers and parts; and even the behavior of the medicine man leading the ceremony. Even limiting the discussion to focus on the actual material instances of eagle and Thunderbird images visually present during the Sun Dance ceremony, one can begin to see the great significance of eagles to Native ceremonial life on the Missouri River.

The center pole of the Sun Dance lodge is central to the Sun Dance ceremony. The center pole is cut fresh from a cottonwood tree. Contemporary Blackfoot elders who have participated in this ceremony noted that finding and cutting the right tree is of utmost importance. They strive to drink the sap that bleeds immediately upon its cutting; the sap comes from the core of the tree, which in turn depicts the shape of a star. This is one of the holiest moments in the ceremony. The center pole is blessed and prayed over as it is prepared for use, and would "transmit the power" of the Sun Dance during the ceremony (Fitzgerald 1991:146). It is usually forked at the top and augmented with branches that represent an eagle's nest (Figures 7.11 and 7.12).

A man with eagle medicine would climb during part of the ceremony and imitate an eagle (Lowie 1915b, 1918; McClintock 1999; Nabokov 1970; Voget 1995). According to Assiniboine accounts of the Sun Dance, an eagle was carved into the center pole near the top to represent the Thunderbird, while a buffalo was carved closer to the foot of the tree (Lowie 1909:61). The appearance of the Thunderbird along with the buffalo, which has a well-known reputation as the most significant animal species in the social, economic, and spiritual life of Plains peoples, reasserts the significance of birds in sustaining this way of life.

Figure 7.12. Crow Sun Dance Lodge c. 1909. Note the bison head and stuffed eagle hanging from the center pole. Photographer unknown, Museum of the Rockies.

CHAPTER EIGHT

Birds as Objects

The best known representations of "the Indian" in North America almost unwaveringly include a Native American in a large feathered headdress, or perhaps with a single feather affixed on the back of his or her head. While these styles are by no means common to all North American Indians, they demonstrate the widespread importance of bird parts in material culture. George Catlin's (1989) and Karl Bodmer's (Hunt et al. 2002) portraits of native people they encountered on their journeys through the Great Plains in the 1830s rarely show a subject without some kind of avian adornment, whether it be feathered headgear, a feathered fan, a staff, shield, jewelry, or other object. Today, the descendants of those men and women whose likenesses Catlin and Bodmer painted examine these portraits to ascertain social and ceremonial identities and affiliations evident in traditional clothing and other gear.

Bird parts were used for utilitarian purposes, as well as to decorate pieces of regalia or ceremonial objects, to denote personal status, identity, or power. Written records from the early encounters between Europeans and Native Americans offer abundant commentary on the use of feathers, plumes, and bird and animal parts in personal adornment and material culture. In fact, the use of feathers and bird parts in the material culture of the Missouri River tribes is best known from descriptions by the first explorers in this region since, other than bones, bird parts do not often persist in the archaeological record, except under exceptional preservation conditions.

Bird bones have the best chance of all parts for preservation in the archaeological record, but the small size of many of these remains may preclude archaeological recovery or, if recovered, their poor preservation makes identification difficult. Bird feathers, quills, and skins are perishable, and therefore do not provide good representations of the use and context of such materials before the accounts given by early explorers. Several archaeological studies carried out in the 1970s focused on avian remains from Plains sites. These made great strides in linking archaeologically recovered avian bones to species represented in the ethnohistoric and ethnographic record (Parmalee 1977a, 1977b, 1979; Ubelaker and Wedel 1975). Combining these with written accounts and contemporary interviews we can see the significance of birds to Missouri River tribes through time.

TRADE

In one of the earliest accounts of the Missouri River tribes, La Vérendrye visited the Mandan in 1738 and commented that "these people dress leather better than do any other tribes, and do very fine work on furs and feathers" (Thwaites 1904:221; see also Blegen 1925). He also noted the importance of trading bird parts within and among tribes, explaining that the Mandan, "knew well how to profit thereby in trading their grain, tobacco, peltries, and *painted feathers*, which they know the Assiliboille [Assiniboine] highly value" and who "had purchased everything which their means permitted, such as painted buffalo-robes; skins of deer and antelope well dressed and ornamented with fur; *bunches of painted feathers*; peltries; wrought garters, *headdresses*, and girdles" from the Mandan (Thwaites 1904:221, emphasis added).

The trade of feathers, agricultural products, and other goods fostered an intricate network among the Mandan, Hidatsa, Arikara, Crow, Assiniboine, and Blackfoot tribes, based on the strengths and wants of each tribe. European traders introduced Old World dyes to the tribes, which became very popular over time. In fact, Zedeño was shown feathers dyed in magenta, violet, and bright green colors as well as packets of unused dye inside a Mandan bundle no longer in use. It is also possible, according to a Blackfoot consultant, that the sacred blue paint used historically in the Blue Thunder tipi and corresponding bundle was trade paint that came to replace naturally occurring green paint. They told how lightning burned down a Green Thunder tipi just before a ceremony. Since then, they have used blue paint instead of green.

Hudson's Bay Company records note that European traders relied on waterfowl and other non-migratory bird species as a food source, as well as for trade (HBCA York Factory Fur Returns, B239 h/1). Goose and partridge feathers; crane, eagle, goose, and swan quills; and swan skins were also frequent trade items across the northern Plains. The Assiniboine served as middlemen between the Europeans and other tribes during the early years of Hudson's Bay Company's York Factory inland trade (Ray 1974:68–69). They acquired European trade goods and passed them on to other Plains tribes at a significant mark-up, exchanging their "used and deteriorated kettles, axes, knives, guns, and other trade goods" for corn, leather, feathers, and other materials less abundant in their territory (Ray 1974:88; Rodnick 1938). Considering that evidence of Assiniboine eagle trapping is rare, trade with the Mandan and Blackfoot became an important way to obtain valued eagle feathers for use in dress, regalia, and ceremonial items (Denig 2000:195; Will and Hyde 1964:179–180).

The Blackfoot had abundant access to Golden Eagles and they were heavily invested in the eagle feather trade. Eagles were more plentiful in the northern Blackfoot territories, where five eagles might be traded for a good horse, compared to farther south where only two eagles could purchase the same item (Grinnell 1920:240). Brings-down-the-Sun "a celebrated medicine man of the north," according to McClintock (1999:312), supported his family through eagle trapping by trading the feathers he acquired to the Piikani in the south, who used them for regalia and ceremonial objects (McClintock 1999:428, 432). One of our Blackfoot (North Peigan) consultants explained that his male ancestors were eagle trappers by occupation, and that the Golden Eagle feathers were prized so highly that one eagle might be traded for a horse. Washington Matthews (1877:28) also reported that the Mandan could trade a single tail feather from a Golden Eagle to another tribe for a "buffalo-horse, i.e., a horse swift enough to outrun a young adult buffalo in the fall."

Other bird parts were also important to the trade relationships of tribes around the Missouri River. One Mandan story mentions that in the past, when the white buffalo robes were very scarce, they were a highly valued trade item. When a visiting tribe came to a village offering one of these robes, the people would bring their most prized items in their possession, among these "*skins of red birds*, buffalo robes tanned, *heads of meadow larks* and dried bears' intestines which they used as ribbons" (Beckwith 1937:107, emphasis added). A red bird is associated with the origin of rainbows (Thwaites 1906:374) and also appears in the Mandan story of "Brown Old Man" as leader of the birds a long time ago when black bears trapped eagles (Bowers 2004:382). Red Bird was captured by Old Black Bear, but explained to the bear that if he was killed there would not be any more birds left. Old Black Bear released him, but birds like Red Bird were never seen again. Red Bird's son, Old Brown Man, could turn himself into a spotted eagle, and brought buffalo and rain to the people (Bowers 2004:376).

These accounts reveal the importance of red birds in Mandan culture and their value as trade items. Stories also tell that the ancient Mandan traded the yellow crescent of the meadowlarks for shell bowls from a [mythological?] tribe they called the Maniga, which lived on the opposite side of the Mississippi where it flows into the ocean (Bowers 2004:132, 156). Meadowlarks were sometimes dropped into the river as people crossed it to calm rough waters. The meadowlark is also valued because it warns war parties of nearby enemies, and it is a predominant character in the Okipa Ceremony (Bowers 2004:132).

Maximilian recorded the trade of Pileated Woodpecker heads to be used on the pipe of the Mandan Adoption Ceremony (Bowers 2004: 329; Thwaites 1906:319). The Pileated Woodpecker's historic range did not reach far above the mouth of the Missouri River, although its range has expanded over the past 40 years (Northern Prairie Wildlife Research Center [NPWRC] 2006). The scarcity of this type of woodpecker added "a considerable expense" to the acquisition of such a pipe, which required the upper bill and distinctive red crown of that bird (Thwaites 1906:319). The bird's heads are associated with Old-Woman and corn, perhaps because they live far to the south where Old-Woman resides. The head of a Pileated Woodpecker had to be brought from St. Louis, and it was traded for a large buffalo robe, horses, and corn (Bowers 2004:330; Thwaites 1906:320). These stories and events strengthen the notion that the Mandan were closely related to Mississippian people and environment.

UTILITARIAN AND RECREATIONAL OBJECTS

During the course of their wandering over the Plains during the summer, the Assiniboine planned their route to include several of the well-known lakes. These places had been known to them for generations as the moulting places for wild geese whose wing feathers were the best ones designed by nature for use on the arrows. When the band would be camped near one of the larger lakes, the older and experienced men and women picked and preserved these feathers for future arrow making. Once collected, the people sorted carefully the feathers, and those which they knew would be usable they packed neatly in the dried lower jaw of the pelican which they then wrapped in buckskin before putting away [Dusenberry 1960:54].

Arrows were a key piece of hunting technology before, and even after, the introduction of firearms to the Missouri River region. Bird feathers were attached to the end of an arrow shaft to stabilize the flight of the arrow and improve accuracy (Figures 8.1 and 8.2). Arrows are considered utilitarian because of their use in everyday life for hunting, but they can also take on sacred significance as part of a medicine bundle. For some tribes, the arrow originated as a gift from the Creator and therefore has a spiritual element innate in the object itself. For example, in one story, arrows were given to the Blackfoot by Old Man Napi when he created the earth. He taught them how to make arrows so that they could protect themselves against the buffalo and taught them how to use the parts of the buffalo (Grinnell 1920:140). The bow, on the other hand, has celestial origins tied to Scarface's journey to see the Sun. Similarly, Old-Man-Coyote "taught the [Crow] to make arrows, taking bird feathers for feathering and stones for arrow points" (Lowie 1918:18). Bow and arrow sets had their own bundle. A Blackfoot consultant told us that in his youth he saw a bow and arrow set inside a Thunder Pipe Bundle.

Several types of feathers were used to fletch arrows for hunting. In the northern Plains, the feathers of waterfowl were most often collected and used for arrows (Dusenberry 1960; Grinnell 1920; McClintock 1999), although eagle and hawk wing feathers were also split and used for arrows by the Assiniboine according to some accounts (MWP 1942:151). One Crow story mentions that "the prairie-chicken wing is always used for arrow feathers. Owl feathers are best" (Lowie 1918:296), but Lowie (1922a:230) does not specify the type of feathers used for arrows in

Figure 8.1. Arrowshafts with feather attached, recovered from Lookout Cave, Montana. Bureau of Land Management Billings Curation Center. Photograph by John Brumley.

Figure 8.2. Cut feathers ready to be used for fletching, recovered from Lookout Cave, Montana. Bureau of Land Management Billings Curation Center. Photograph by John Brumley.

his later discussion of arrow-making among the Crow. He comments that arrows were crafted by experts and that 10 high-quality arrows might be valued as equal to a horse.

For the Hidatsa, Wilson (1911) was told by Wolf Chief that prairie-chicken feathers were *never* used to feather arrows because they were so difficult to find in tall grass, "for prairie-chickens are adept at hiding in the long grass." A later exchange between Wilson (1924:162) and Wolf Chief suggested that the arrows he used for enemies were always feathered with prairie-chicken feathers because Wolf Chief's father, an arrow maker, said they made the arrows fly the fastest. This may suggest that arrows were feathered differently for warfare and hunting purposes. Further, Wolf Chief explained to Wilson (1911:45) that

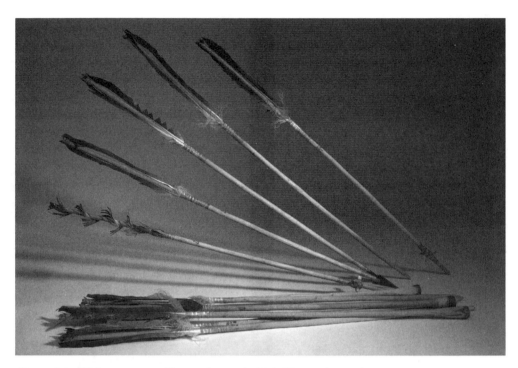

Figure 8.3. Hidatsa arrows (date unknown). AMNH, 50.1/6014 C-L.

when feathering an arrow in his youth, the feathers of small birds were split and scraped while holding one end of the feather in one's mouth. But when eagle feathers were used on arrows, the feathers were not held in one's mouth because the Hidatsa "thought eagle feathers were magic and might poison us if put in the mouth" (Wilson 1911:45). In this case, the eagle feather was placed under the toe of a moccasin and the surface of the quill end was scraped or ground until smooth with scoria (burnt lignite), a stone that can often be found floating down the Missouri River. When enough feathers had been prepared, three were attached to the arrow with sinew and the feathers trimmed (Wilson 1911:46–47) (Figure 8.3).

Bird feathers, bones, and parts were also modified for other uses as tools, utensils, everyday clothing or other utilitarian items (Culin 1975:174). As noted previously, the Assiniboine used the dried jaw of a pelican as a storage container. Molted feathers from a gull-like bird frequently found on the Missouri River were collected from the riverbanks. These feathers could be used for quillwork after removing the quill, one of the oldest adornment methods (Wilson 1909:130). Bird quills were considered superior to porcupine quills, and also lasted longer (Wilson 1911:289–290). Wilson purchased an awl case decorated with these feathers, which he was told were selected specifically for making the ornaments that went on the front of a tent

(Wilson 1909:130). Animal skins were occasionally made into caps by the Blackfoot in the winter. These might be made from large birds such as the sage-hen, duck, owl, or swan, or the skins of other small animals like a badger or coyote (Grinnell 1920:196).

Bird parts and feathers were also common in certain recreational activities. There are eight dice made of crow claws in the Assiniboine collections at the Field Museum of Natural History (VanStone 1996:14, 59, fig. 28e). Wolf Chief of the Hidatsa recalled using duck feathers and owl feathers on wooden arrows that were used to hunt small game and for playing games (Wilson 1924:162). In the hoop game, which is widespread throughout the Missouri River region, hoops may be decorated with feathers and other adornment such as quills and metal trinkets (Gough and Brown 1988:375; Lowie 1909).

REGALIA

When Maximilian encountered a camp of Mandan, Hidatsa, and Crow on the shores of the Missouri near Fort Clark in what is now North Dakota, he commented on the physical appearance and attire of each group. Generally, he said, "In dress, these three nations differ little. . . . On their heads they wear as many feathers as enemies they've slain, in natural colors or with yellowish [or] reddish

Figure 8.4. Left: Blackfoot straight-up bonnet with double trailer of eagle feathers, ca. 1880s. Source: Scriver (1990:49). *Right*: Blackfoot split-horn bonnet with porcupine-quilled feathers tipped with hawk bells, ca. 1840s. Source: Scriver (1990:53).

brown transverse stripes. These feathers often hang down in back like a fan" (Witte and Gallagher 2008:196). Maximilian also describes the Hidatsa's dress as "especially beautiful and richly embroidered with beads and feather quills of the most vivid colors, with large numbers of braids and strings of leather and hair. The Crows are said to outdo even them in this respect" (Witte and Gallagher 2008:202–203).

The term "regalia" is meant to describe clothing, accessories, and ornamentation that are worn for a particular purpose or event other than ordinary daily attire, such as items worn during a war party, a dance, or a ceremony (Figure 8.4). Feathers, skins, and other bird parts were often incorporated into regalia to express achievements or status, identity, membership in a particular clan or society, a personal relationship with a certain spirit bird, or for the specific purposes of a dance or ceremony. By donning the feathers, skins, or other parts of a bird, or by painting oneself to represent a certain bird, a person might take on the power of that bird's spirit. Eagles, hawks, owls, crows, ravens, ducks, geese, swans, gulls, cranes, turkeys, roosters, grouse, magpies, and meadowlarks are all represented in historical accounts of regalia for the Missouri River tribes, indicating the diversity of birds present in the region and their incorporation into the life of these tribes.

The distinction between regalia and ceremonial objects is rather vague. Regalia may be part of a complex collective of ceremonial items, which are all interrelated as part of an animate sacred bundle. This section focuses on those items that are worn on the body or as part of a ceremonial outfit, representing a physical, embodied connection between the individual and the spiritual power of bird feathers and parts worn during a ceremony. This is differentiated from ceremonial objects such as pipes, whistles, drums, bundles, effigies, and other objects that are used in ceremonies, but not worn by participants.

Some descriptions of regalia only refer to feathers and bird parts in the most general sense, without identifying the species from which they came, or even the color of the feathers. This likely owes to the fact that many of the early ethnographers were more interested in other topics and often did not have the ornithological knowledge or concern to identify feathers or skins to bird species. In other cases, the specific kind of feather may not have been relevant to its use, or may have been lost in translation. For example, when the Hidatsa culture hero Charred Body prepared to woo a girl, he fixed his hair into a pompadour and "stuck a plume in at the place where he tied it up" (Beckwith 1937:26). In this case, the type of feather is not identified and may not really be important. However, Lowie's (1909:70) description of the regalia for the Assiniboine no-flight dance (*nampe'c owa'tc*) is similarly vague: "rawhide rattles decorated with feathers and flannel, two spears wrapped with otter-skin, and buffalo-horn headdresses." It is likely that a specific type of feather was desirable to decorate the rawhide rattles, but we cannot tell from the general level of description in the source.

Lowie's descriptions of Assiniboine material culture are equally general and only identify feathers specifically when they come from the eagle. More precise bird identifications are made in the "Mythology" section. These do not rely on the ornithological identification skills of the ethnographer, but rather the knowledge of the storyteller and the ability of the interpreter to translate bird names into English. Therefore, it is very possible that the tribes of the Missouri River used a wider variety of bird species in regalia than is mentioned in ethnographies. Despite this, they noted the abundance of feathers in regalia, with feathers and bird parts appearing on clothing, robes, war shirts, headdresses and headgear, shields, lances, arrows, jewelry, and even adorning horses as they went into battle. Regalia, ultimately, may refer to social and political standing, either in everyday life or within the context of social and ceremonial occasions.

Eagles

In the olden days there was a young man who had great powers and was a type of sacred person—even greater than a medicine man. He was adopted by both the buffalo and the eagle, and they have given him great, great medicine.... The young man lived with the seven bulls and learned from them and from the eagle. The eagle gave the young man the power of the eagle and the special power of the eagle plume. The eagle plumes are to be found on the breast of the eagle and are sometimes called "breath feathers." We use eagle plumes in all of our ceremonies, and they have great power. To have the power of the eagle plume is a great thing. You remember that Grandma also had the yellow eagle plume as her medicine and carried a yellow eagle plume with her wherever she traveled. The young boy had this same protection, and it proved to be great medicine [Fitzgerald 1991:110–111].

Because of the power they convey, eagle feathers are by far the most widely employed and documented bird part used in regalia. The feathers, claws, and other parts of eagles are featured in costumes for dances and ceremonies (Figure 8.5). Eagle wings are carried as a "mark of distinction" during many tribal ceremonies (Linderman 2002:32). In traditional times, performances and ceremonies using eagle parts had varying levels of social and religious significance. Sometimes eagle feathers were worn in social courting dances, like the Crow and Hidatsa Goose Egg Dance where dancers wore a flannel headband with two eagle-tail feathers tied to it (Lowie 1924:361–362). In other settings, such as the Sun Dance ceremony, eagle regalia represents a connection to the eagle as the Sun's messenger, and participants and dancers wearing the eagle plumes might evoke the eagle by wearing its feathers and imitating the behavior of an eagle (see McClintock 1999:310 for one vivid example).

The ceremonies observed and recorded by early ethnographers are by no means exhaustive of the variety of dances, performances, and ceremonies that were carried out by the Missouri River tribes. The accounts of the regalia used in these contexts are likewise incomplete. Yet, there are more than 25 ceremonies recorded historically and ethnographically that include specific eagle feathers and parts in regalia. Examples from each tribe furnish an overview of the pervasiveness and variety of eagle parts used in regalia. Due to the limited access to eagle parts under the eagle protection legislation, eagle feathers and parts are more likely to be used in sacred than social contexts today (Murray 2011). Eagle feathers are most commonly worn or held by men, although on certain occasions a woman might wear an eagle feather or headdress or carry an eagle fan in a ceremonial context

Figure 8.5. Eagle feather fan from the Fort Berthold Indian Reservation (date unknown). AMNH, Division of Anthropology, 50.1/5411.

(Kennedy 1961). The conventions for the handling of eagle feathers varies with each of the tribes.

Assiniboine dancers donned fox-teeth headdresses adorned with four war eagle feathers for the Fox Dance (*To-kah-nah Wah-Che*) (Denig 2000:167). They also attached Golden Eagle feathers, some dyed green, to each horn of an Assiniboine buffalo headdress (VanStone 1996:11). For their Grass Dance they attached eagle tail feathers to sticks used by the dancers to signify the number of war bonnets he gave away (MWP 1942:194). Dancers in the Blackfoot Kit-Fox (*Sinopaix*) Dance wore multi-colored eagle feathers with weasel skins attached on either side of the feather (McClintock 1999:447). Eagle feathers are also an important part of the robe used during the Blackfoot Sun Dance, which was given to Brings-Down-the-Sun in a dream (McClintock 1999:519). Eagle plumes are the "most visible adornments" on each Crow dancer in the Sun Dance (Fitzgerald 1991:159), and members of the Eagle chapter of the Crow Tobacco Society wore eagle feathers, moccasins with eagle claws attached, and eagle thumb claw wristlets (Lowie 1919:130–131).

Tiers of eagle feathers adorn the buffalo horn headdresses worn by Arikara Young Buffalo Society dancers (Mails 1973), and during the Mother Corn ceremony, the woman who represents Mother Corn wears two eagle down feathers (Will 1930). The regalia of holy men from the Mandan Coarse Hair Society included feathers from birds of prey, worn according to one's personal experiences in war (Mails 1973:187). Eagle down and tail feathers were worn on headbands by dancers in the Enemy Women's Society (Mails 1973:179). Eagle feathers were attached to the backs of girls' heads when they danced for the Hidatsa Skunk Society to honor a successful war party, although they hung downward in contrast to the way a plume was worn in a man's hair (Gilman and Schneider 1987:106; Peters 1995:73; Wilson 1911:176). The Hidatsa Crazy Dog Society's regalia included a sash made with downy eagle feathers that had been dyed red and attached to the tips of eagle feathers (Gilman and Schneider 1987:92).

Eagle feathers as a part of regalia or other ceremonial objects are treated with reverence and care in respect for the power carried in an eagle feather. Handling and wearing eagle feathers is often strictly proscribed and only practiced by those who have eagle power and handling rights. Further, maintenance of regalia that includes eagle feather or parts requires special permissions and procedures, especially in ceremonial contexts. For example, if a feather falls off a piece of regalia during a ceremony, it may require a shift in the entire ceremony to address and mitigate the potential effects.

At powwows, if an eagle feather drops and hits the ground and somebody sees it, they dance around it until that song is over. Then the powwow committee or the announcer will ask a veteran of war to come and pick that feather up. To me, if I were to drop it, I'd have someone come get it, a veteran of war—don't matter what war it is. Then turn and give something to the committee, and say hey, my apologies for not checking my outfit out, not securing everything down. It's my bad, because you are respecting this great feather that Native Americans love and want so much to get a hold of. They ain't easy to come by these days.

—Mandan-Hidatsa consultant

Ethnohistory indicates that this tradition has some antiquity. McClintock (1999:273–274) observed during the Blackfoot Okan ceremonies that when an eagle feather fell out of Sepe-nama's war bonnet during the Kisapa Society dance, he chose another noted warrior, Bear Chief, to pick up the feather for him following specific protocol. For Sepe-nama to pick it up himself would bring bad luck. The ceremony shifted to address the fallen eagle feather, with the dancers circling the feather in single file three times before Bear Chief picked up the eagle feather on the fourth circle. After this, everyone took their seats and the ceremony continued.

One Mandan-Hidatsa consultant explained that for powwows, young children are given less valuable feathers that were easier to attain than eagle feathers, in contrast to older dancers who knew how to properly handle feathers. Young generations get Red-tailed Hawk feathers and pelican feathers, "just to get them started" with learning how to take care of feathers. More experienced dancers can wear eagle feathers in their regalia because they know how to respect and care for these feathers.

In addition to the incorporation of eagle parts and feathers into regalia, eagle feathers were commonly used in the past to "count coup" or connote important deeds or bravery on the battlefield such as being the first to strike an enemy. Although a variety of bird feathers might be awarded for counting coup based on the specific act of bravery or the tribe one belonged to, eagle feathers, and especially those from the golden eagle, were often reserved for the highest honors. The way a warrior wore the feather, if the feather was dyed a certain color, or worn with other materials such as horsehair, all communicated information about a one's warrior status and act of bravery.

These feathers were worn to signify warrior status in a ceremony or dance, or they were worn into battle. Members of their community (and their enemies) understood what the different feathers and styles of decoration signified. These "honor marks" were important and easy-to-read indications of individual accomplishment and status (Weitzner 1979:293). Hidatsa women would also receive honor marks, given to her by her clan aunt, for taking good care of her gardens and the earth lodge. These were worn

on a deerskin belt adorned with feathers, although the type of bird feather used is not mentioned (Weitzner 1979:301).

Weitzner (1979) describes in detail the variety of ways an eagle feather might be decorated and worn by a warrior to convey to the community a certain event when he returned from a war party. The intricate signaling of eagle feathers demonstrates the high respect people hold for the eagle and the depth of integration of this bird into the social and religious structure of Plains tribes. Wing feathers and tail feathers communicate different events, as do other animal parts worn in association with the feather. For example, in the Hidatsa tradition, a tail or wing feather worn on the arm of a war party leader meant they had found or trapped an eagle, while an eagle feather worn next to white horsehair dyed red signified that the leader had captured one or more horses (Weitzner 1979:292–293). A middle eagle-tail feather worn in the center of the back scalp lock meant that a warrior had counted first coup on the enemy, and an eagle wing represented less significant strikes on the enemy when worn in a hair braid.

Eagle feathers continue to hold significance for honoring brave acts, even though access to eagle feathers has been greatly diminished because of legislation protecting eagles and eagle parts and regulating their circulation. Only enrolled members of federally recognized tribes can obtain a permit from the U.S. Fish & Wildlife Service to possess eagle feathers for religious purposes. The high-demand for a limited supply of eagle feathers and parts has created a long waiting list. Applicants may have to wait as long as three and a half years to obtain the eagle parts they request and, as previously mentioned, this process disconnects Native people from the spiritual process of eagle trapping and also changes the "intention behind the feather" or other eagle part (Murray 2011:148). Furthermore, eagle feathers and parts may not be "sold, purchased, [or] bartered" to another person, as was done in the past, but only passed down through family lines, generations, or from one Native person to another for religious purposes (16 U.S.C. 668–668d). Despite these constraints to acquiring eagle parts and feathers for traditional uses, the objects remain important to maintaining tribal identity and traditional cultural practices.

Modern day warriors, those Native men and women who serve in the armed forces, are still honored with feathers for their military service and they may wear them as regalia (Murray 2009, 2011). Feathers are also given to those deployed in the armed services to wear as protection from harm through the eagle medicine. For example, Yellowtail, a Crow medicine man, sent a Sun Dance eagle feather with his adopted grandson when he went off to the Vietnam war in 1969. With the restrictions placed on acquiring eagle feathers, however, it is likely that other feathers might be substituted for this purpose, although they would not hold the same meaning or power as an eagle feather. Eagle feathers are also still integral to many ceremonies as a part of regalia and ceremonial objects.

Eagle feathers are given as a gift from one person to another to honor someone, or to recognize their potential. One Mandan-Hidatsa informant told a story about how he was given an eagle feather as a gift after playing his guitar for another man:

> If I'm not at college or working, the majority of my time was spent writing music. I write my own music and play my guitar. I went out to Yakima, Washington, and I had all these songs and my uncle said, "Hey I outta take you out to this old fancy-dancer from Oklahoma. He wants to hear your music 'cause I told him about you." I went out there and jammed. I must have sang for an hour and a half, and he comes up to me and puts a shoe box right here on me. And I never thought, well I thought he was giving me a pair of tennis shoes or sandals or something. He told me to open it up, and there was all these eagle feathers in there, from the wing all the way to the tail. He says, "Take one. Something's telling me you are going to go very far with that guitar and your songs. It may not be right away, but ten or twenty years from now it's going to happen. I want to reward you with something."
>
> I said, "Hey man, thank you. I just met you an hour ago!" He said, "Yeah, but there's something about you." We have these feelings, everybody has these kind of feelings. So I started digging through them feathers, and I laid them all out. Then there was this really pitiful one. It wasn't up to par. It had been roughed up, banged up, you know? Just a little plume, and I took it out. I said, "Grandfather, thank you for giving me this feather."
>
> He looked at me and said, "Why are you taking that one? There's a lot of nicer ones in there." I told him it's not about the look. You are already judging these when you judged it as being a bad, pitiful one. So I take that consideration a very big compliment, because it was at the bottom of the stack. Who was going to find him but me? So there was a connection right there with me and that feather. So I took it and I used it, I cleaned it up. Some of us go through hardships. . . . I used that as a connection with that plume. I could have had a nice feather, but something caught me with that feather.
>
> —Mandan-Hidatsa consultant

Hawks and Falcons

Hawk feathers were sometimes used in ways similar to eagle feathers and in some cases could be substituted for eagle feathers because they have similar markings and significance (Schaeffer 1950:40). A consultant from Fort Berthold said that feathers from the Cooper's Hawk could be used if eagle feathers were not available, but they were regarded as less sacred than feathers from an eagle. The Blackfoot word for the Rough-legged Hawk is *Ikpaksoát-simiop*, which translates to "resembles eagle tail feathers" or in some accounts, "little eagle" (Schaeffer 1950:40; Hungry-Wolf 2006:135). The tail feathers from the Rough-legged Hawk could be substituted for eagle feathers, especially in the winter when eagles became more scarce during their migrations (Hungry-Wolf 2006:135; Sterry and Small 2009).

The hawk features prominently in the regalia of many ceremonies, and especially those that relate to bison and bison hunting rites. For example, buffalo-calling rites were given to the Mandan by a "small hawk," who, "being a good hunter, taught the Buffalo-calling rites," along with rain-calling rites to the people (Bowers 2004:271). The Small Hawk ceremony is performed by Small Hawk bundle-owners in the Okipa ceremony, one of the most important events of the year, as a part of the buffalo-calling rites. Historically, the leader of the Buffalo-calling ceremony was a Small Hawk Bundle holder who officiated as the singer for the ceremony, and as a part of his regalia he wore a hawk on his head, while his wife wore magpie feathers (Bowers 2004:272). Members of the White Buffalo Society of the Mandan also wore a fan of hawk feathers as part of their headdresses (Figure 8.6; Mails 1973:179).

The Hidatsa Lumpwood Society held close association with buffalo. Its dancers had two eagle feathers attached to their headdresses, and the leader's club was decorated at its head with eight hawk feathers (Mails 1973:167). Hawk feathers also appear in regalia for other dances that are not related to buffalo-calling, such as the Arikara Hopping Society dance, which uses rattles decorated with hawk feathers, and the Hidatsa Dog Society, which utilizes hawk (or turkey) feathers to create the fan-like back of their headdresses (Mails 1973:150, 167).

Hawk feathers were also used to denote war honors. An account from the Fort Belknap Indian Reservation dating to the mid-twentieth century recalls that the Assiniboine warriors used hawk feathers to indicate their war honors before eagle feathers were introduced by the Dakota Sioux into the northeastern Plains (Dusenberry 1960:59). The number of coups made by a warrior was denoted in

Figure 8.6. Hawk feathers used in the headband of the White Buffalo Society of the Mandan, along with skunk fur, c. 1910. AMNH 50.1/ 4315.

notches made on a single feather, and the kind of feather further indicated the nature of the deed. For example, a sparrowhawk (American Kestrel) feather could indicate that the wearer has killed an enemy singlehandedly, while a feather from a different bird would represent a different act (Dusenberry 1960:59).

In addition to serving as personal marks of honor, hawk feathers were also incorporated into the regalia and medicine objects men brought with them into battle. One example of this is the use of hawk wing feathers by Hidatsa men to communicate that they were about to lead a war party. The man planning to lead the party would hang several bunches of five hawk wing feathers on the right-hand post of the entrance to his earthlodge. These feathers would then be worn by young Hidatsa warriors who lacked protection in battle because they had not yet received a vision from a spirit bird or other spirit animal. The hawk feathers were incorporated into the young men's war regalia to ensure their safety. This service performed by the war party leader required compensation (Weitzner 1979:289).

Ravens

Raven feathers are a very important part of the Blackfoot Okan, who trace their roots in the ceremony back to the origin of its Medicine Lodge. Recall that the ceremony was given to the Blackfoot culture hero Scarface by the Sun in return for saving the Sun's son, Morning Star, from the

attack of vicious cranes (Grinnell 1913:103–106; McClintock 1999:498). The Sun instructed Scarface on how to build the Medicine Lodge and gave him two raven feathers, explaining that if a man was sick, his wife could build a Medicine Lodge to ask the Sun to heal him. The two raven feathers are an important piece of regalia to be worn by the ailing man during the ceremony (Grinnell 1913:104).

Raven feathers also figure prominently in the Mandan Okipa ceremony. The Okipa, which commemorates many of the Mandan origin stories and includes the rites for buffalo-calling and rain-making, is performed over the course of four days. Lone Man plays a significant role in the ceremony. On the first day, He who impersonates Lone Man opens the ceremonial lodge for the buffalo dancers, wearing a stuffed raven tied to his headdress as part of his regalia (Bowers 2004:125). During the Bull Dance on the third day of the Okipa, the impersonator wears a cap of raven feathers and carries a staff adorned with alternating rows of raven and swan feathers (Bowers 2004:139–140). The raven and swan feathers on the staff are connected to the origin story in which a raven and three white swans help direct Lone Man northward to Dog Den Butte, where Hoita (a speckled eagle) had imprisoned all the animals (Bowers 2004:349). In a related Hidatsa ceremony called the Imitating Buffalo ceremony, which is also closely associated with the Earth-naming bundles, the singer of the ceremony wore a raven feather to represent Raven

Necklace and the role he played in the Earth-naming myth (Bowers 1992:445).

In other ceremonies, raven feathers and parts were also important to regalia. The Hidatsa had a society named for Raven that disappeared very early after European contact, due to loss of knowledge and members because of wars and smallpox (Mails 1973:173). In 1913, the last remaining member of the Raven Society was seen wearing a necklace made of raven skin, although in earlier times a raven skin headband was more common. Another society carrying the Raven as its namesake, the Arikara Shin Raven Society, received its name because raven feathers hung from the "slits" in the young men's shins (Mails 1973:145). Raven parts are also used in the Mandan Crazy Dog Society regalia (Mails 1973:185), raven skin necklaces are worn by the Hidatsa Kit-Fox Society (Mails 1973:271), and split raven feathers are worn for the Sunset Wolf ceremony (Bowers 1992:441). Ravens and raven parts were also worn by Blackfoot warriors with Raven medicine (Grinnell 1920:261) and, along with a coyote tail, by Mandan warriors whenever they were the first to see the enemy (Bowers 2004:72).

Crows

Due to their similar stark black appearance and behavior, crows and ravens are often viewed as closely related by the tribes in the Missouri River Basin. The Blackfoot word for a crow is *maistó*; and for the word "raven," the word for "crow" is expanded with the prefix *Umaχkaisto*, or "big *maistó*" to account for its similar features yet notably larger size (Schaeffer 1950:43). Sometimes crow and raven parts could be used interchangeably in regalia. More often, however, the two species were differentiated, as each held unique functions.

McGowan (1942:100) asserts that the plumage of crows was among the most widely used feathers for the regalia of the Arikara, commenting that "[t]he head and body of the crow was made into a head dress. The tail was also used as an ornament in various ways." A few examples include the rattler of the Fox Society who wore the whole skin of a crow on his head with the tail sticking out from the back (Lowie 1915a:667); the Dance of Youngest Child, in which a piece of swanskin and a crow's feather were worn on the back of a dancer's head (McGowan 1942:118); and the Crazy Horse Society dancers who wore crow feathers in their hair and carried bows adorned with both crow and eagle feathers (Mails 1973:149).

The Assiniboine Crow Society, a war society that restricted its membership to young warriors, made liberal use of crow feathers in its regalia. Ravens and crows were often associated with war, in part because they were always seen at battlefields (Mails 1973:243), but also for the skills of the birds. In the Assiniboine Crow Dance (*Congghai Wah-che*), "[n]eck and head dresses of crow skins taken off the bird entire with wings and head on are worn by all, and crow feathers adorn their lances, shields, and other war implements" (Denig 2000:170).

Similarly, Arikara Crow Society dancers carried lances decorated with crow and eagle feathers (Mails 1973:149). Crow feathers were also used to designate war honors; for Hidatsa warriors who "struck a woman" during a war party, crow wing feathers were stripped in a particular way, attached to a stick that represented "the ancient coup stick," and worn in a headdress made with porcupine quills (Weitzner 1979:296).

Owls

As explained in earlier chapters, owls walk a thin line between life and death, good and bad luck, positive and negative associations. Their powers are both feared and revered. An owl's power is used in healing and, to a lesser extent, in warfare, fortune-telling, gambling, hunting, and guiding. Therefore, the use of owl feathers and parts in regalia is most often associated with societies devoted to healing and medicine, or those that are associated with the spirits of the dead.

The Arikara Owl Society is composed of medicine men who heal others using the power of their namesake (Figure 8.7). Owl Society members wore an owl feather in the hair when they held their meetings, participated in ceremonies or performances, or treated patients (Gilmore 1932). The Arikara believed the feather to be "potent as an aid to drive away the disease" (Gilmore 1932:45–46). The spirit of the owl itself might be called on to draw the sickness out of someone (see Fitzgerald 1991:70–71 for an example using the Snowy Owl). But owl parts are also alive with the spiritual power of owls and therefore can function as healing materials themselves. In addition to wearing an owl feather, each member of the Owl Society also wore as part of their regalia a pair of bracelets made from the skin of an owl's legs, with the feet attached and hanging as pendants from the bracelets. Owl materials are also used in Arikara ceremonies in connection with the spirits of the dead. In the Dance of the Spirits, the Ghost Dancers wear a large cap of owl feathers that hang down their backs. Some carry owl skins (Lowie 1915a:652; McGowan 1942:120).

Owl feathers may also be worn in war regalia, although their context varies from tribe to tribe. Dancers for the Arikara warrior society, the *Taro-xpa*, wear a switch with owl and eagle feathers attached to it. Owl, swan, and crow

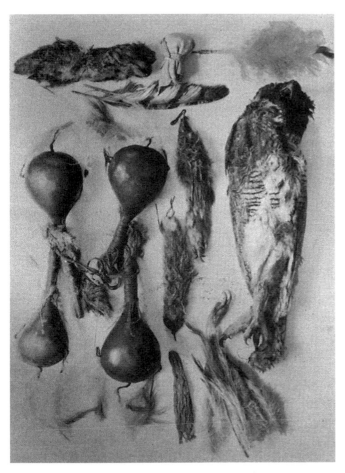

Figure 8.7. A bundle from the Arikara Owl Society. The bundle includes a stuffed owl (*right*), four deerskin rattles (*left*), an owl head, red cedar incense, an eagle down feather and two owl plumes (*top*), and bracelets made from the skin of owls' legs with talons attached (date unknown). Source: Gilmore (1932: Plate IV).

feathers also adorn the society's two lances, because these birds "wished to be the lances" for the society (Lowie 1915a:665). Owl feathers were also applied to the scalp locks of Blackfoot warriors (Grinnell 1920 197), although according to Assiniboine traditions they were only worn by novice warriors (Denig 2000:160).

Owl parts also play a large role in the regalia and cere-monial objects of the various Hidatsa Dog Societies, which included Little Dogs, Crazy Dogs, Dogs, and Old Dogs (Bowers 1992; Gilman and Schneider 1987; Mails 1973). These societies were related but differentiated by the age of the members. Owl feathers were given to these societies by Yellow Dog, who originated the Dog Societies, for them to wear while dancing (Bowers 1992: 196). Owl feathers adorned the headdresses of the officers and dancers. Owl feathers were worn in bunches on the backs of members of the Little Dogs Society and were considered sacred (Mails 1973:171–172). Gilman and Schneider (1987) suggest that magpie feathers might be substituted in the Dog Society dancers' headdresses. Owl feathers in association with a bag of Dog Medicine given to Bug Woman appeared in another Hidatsa ceremony called the Imitating Buffalo ceremony. The dog medicine bag was decorated with owl claws,[1] and Bug Woman's daughter, Rattles Medicine, wore a stuffed owl on her belt for "good health and long life" (Bowers 1992:444).

In a humorous story, a consultant from Fort Berthold told the authors that when he was young he wanted "to be different" and made an owl bonnet for himself to be worn in a Pow-wow dance. His mother was horrified and made him discard the bonnet because of the spiritual danger posed by owls.

Magpies

The Hidatsa consider magpies to be sacred and prohibit their consumption, but magpies may be shot for their feathers (Weitzner 1979:197). Despite their sacred desig-nation, magpie feathers are used for regalia in both formal and informal contexts. Young boys wore magpie feathers in an informal Hidatsa dancing society called Magpie. Although the society was informal, it still held spiritual associations, as the leader of the Magpies who trained the young boys was an older man who held the Woman Above Bundle (Bowers 1992:180). Woman Above was believed to wear magpie feathers. In the Holy Woman ceremony, which is also associated with Woman Above, the women who represent Holy Women are known as magpies or "people from the groves" because magpies are known to live in ash and oak groves (Bowers 1992:325). Magpie Society dancers wear two magpie feathers in their hair as part of their regalia.

Magpie feathers also represent membership in the Holy Woman Society for the Mandan, but their use seems to be more restricted to women (Bowers 2004). Magpie feathers were a part of the Small Hawk Bundle, which could be divided into two sections between a husband and wife, the woman's portion containing the magpie and magpie tail-feathers among other items (Bowers 2004:270–271).[2]

---

[1] Bowers identifies the claws as either "owl or hawk" although the rela-tionship between Dog Medicine and owls, as discussed, makes the owl claw seem more likely.

[2] A man of the Mandan's WaxikEna clan could possess the part of the Small Hawk Bundle consisting of the magpie and its tail feathers, a hawk, sage, hide rattle, pipe, and buffalo skull, while the wife held the sacred robe, gourd rattle, and sage (Bowers 2004:270).

Women with bundle rights could wear magpie feathers in their hair as part of their regalia and carry the ash digging stick at the close of the Okipa ceremony and during the Buffalo-calling rites.

Magpie feathers are included in the headdress for dancers in the Hidatsa Dog Society, although Gilman and Schneider suggest that these can be interchanged with owl feathers (Gilman and Schneider 1987:93; Mails 1973:170). Magpie feathers also decorate the bows used by the officers of the Half-Shaved Head Society, which originated with the Crow Indians (Mails 1973:170).

Waterbirds

The feathers and parts of waterbirds such as ducks, geese, swans, gulls, and cranes are used in regalia for a variety of ceremonies, ranging from the agricultural and buffalo-calling rites of the Mandan Goose Society, to the Big Foot rites of the Arikara Medicine Fraternity, to war honors, and to the adornment of a Mandan "dandy." The scope of uses for waterbird feathers and parts reflects the unfailing presence of these birds in the region and the diverse roles they take on in the lives of the Missouri River tribes.

Goose parts were especially important to the regalia of the Goose Society, which was composed of Mandan women who carried out feasts, dances, and offerings for the waterbirds as they migrated south in the Fall to the place where Old-Woman-Who-Never-Dies lives. The Goose Society also ensured the waterbirds' return north in the Spring, signaling the planting season. These practices originated when a Goose spoke to the Mandan and told them, "I will go to the edge of the big rivers. When it is time for you to prepare something for me to eat, I shall return. When I shall have come back, you may proceed with your garden work, and you will be sure of success" (Mails 1973:150).

Among the Mandan, Good-Furred-Robe founded the Goose Society and set out the appropriate regalia for ceremonies (Bowers 2004). Members of the Goose Society wore the rolled skin and the head of a goose as headbands, and two young girls in the society had their mouths painted, one blue and one black, to represent the geese (Bowers 2004:196). It appears that the materials used for the Goose Society headbands were somewhat malleable to allow for the use of other waterbirds, as the specimen featured in Gilman and Schneider (1987:33) was made of a duck beak, duck skin, and green neck feathers.

According to McGowan (1942), swan feathers were held in high esteem for use in the headdresses of the Mandan. The use of swan feathers in Lone Man's war-bonnet trailer during the Bull Dance on the third day of the Okipa ceremony is significantly linked to a Mandan origin story, where a raven and three white swans helped to direct Lone Man toward Dog Den Butte to free the animals that were imprisoned there by Hoita (Bowers 2004:139–140, 349). Raven and swan feathers are also attached to Lone Man's staff in alternating rows. Because the swan and raven are so closely associated with Lone Man for the Mandan, the two men who represented swans in the ceremony for the third day of the Okipa were required to have Lone Man rights (Bowers 2003:131). The North Peigan story of "The Horns and the Matoki," which are the most powerful esoteric societies of the Blackfoot, states that swan skins used in the two bonnets worn by the Horn Society during the Buffalo Dance were originally given by the bison in dreams to Calf, Cow, and Black-Crow (Wissler and Duvall 1908:120).

The context and rights of use for swan feathers may have been different outside of the ceremonial circle. For example, Catlin described that Mandan dandies or "beaus," that is, men who did not like to concern themselves with typical male activities such as hunting and war, often adorned themselves with feathers from birds that were easy to acquire, such as swans, ducks, and turkeys (Catlin 1989:111). Further, swan feathers were also used as marks of honor for Mandan warriors; the first man who reported spotting an enemy tipi on a war raid had the privilege of wearing the white wing feathers of a swan in his hair (Bowers 2004:73).

The incorporation of bird parts and feathers into regalia may be used to evoke the spirit of that bird and its power during a ceremony. For example, the Duck Order of the Arikara Medicine fraternity makes use of duck-bills in its regalia, strung onto otter-skin.[1] During the Big Foot portion of the Medicine Ceremony, those wearing the duck-bill necklaces would bring rushes from the river into the lodge to represent the marshes where ducks nest:

> A second virtuous woman would be given their necklaces, which she would hold out to the north and then throw among the rushes. Immediately a sound would be heard like the quacking of the spirit duck, which was believed to dwell in the body of the leader. When the fire died down, he would pass through it to the opposite side of the fireplace, imitating as he did so

---

[1] The Duck Order is also called the Big Foot Order, referring to the direct translation of the Arikara word for 'duck' [*AxwahAtkúsu'* (Parks 1986:21) or *Hwat-kúsu* (Curtis 1909:64)] into English.

the actions of the duck. Observers heard the sound of a large flock of ducks quacking and even the flapping of their wings against the surface of the water [Mails 1973:153].

The power of the duck-bill necklace as a piece of regalia and an important ceremonial object is affirmed by the appearance of the duck spirit during the Big Foot ceremony. Other waterbirds are also important to regalia. For example, the feathers of a Ring-billed Gull (*Larus delawarensis*) were worn by Assiniboine dancers to signify a certain warrior status. Their White Crane Dance (*Pai-hun-ghe-nah-Wah-che*) did not actually use physical materials from a crane in its regalia, but featured a two-headed crane painted onto the elk-skin robe worn by the leader (Denig 2000:169).

Gallinaceous Birds

Domestic and wild fowl such as roosters, turkeys, and grouse also appear in regalia. It is important to note that chickens and roosters were not native to North America before the arrival of Europeans so their incorporation into regalia not only helps us to estimate the temporality of their use, but also demonstrates the adaptability of tribes to integrating non-native birds into ceremonial items. A good example is the Assiniboine incorporation of rooster feathers into a feathered belt worn by dancers in the Grass Dance (MWP 1942:197). The rooster feathers are arranged into bustles to resemble a rooster's tail, representative of a vision given to a member of the Sioux by a "very large white rooster" (MWP 1942:193). It is unlikely that a rooster could be mistaken for any other kind of gallinaceous bird indigenous to North America, so it appears that this dance was a relatively recent acquisition by the Sioux, which was then transferred to the Assiniboine.

Wild Turkeys (*Meleagris gallopavo*) appear in the regalia of several Missouri River tribes, although the specific significance of turkeys in native cosmologies is not evident in the sources available. Dakota Sioux, bands of whom came to live along the lower Missouri River in the historic period, considered the turkey to be the original "eagle" at the time they were still living in the forests of the upper Mississippi region (Zedeño et al. 2001), but it

may not apply to Siouan speakers of the northern Plains. According to Dorsey's accounts of the Arikara, turkeys, along with owls, woodpeckers, and snakes, were regarded as mysterious creatures, and they were associated with the witchcraft of people living to the south (Dorsey 1904:22). Alternatively, turkey feathers have replaced eagle feathers in many social dances for the Crow because of the difficulty of acquiring the latter after eagles became a protected species (Voget 1995:175). The significance of turkeys and turkey feathers may vary according to tribe.

Turkey feathers were utilized in the Arikara Hot Dance. Dancers arranged the feathers in their hair to resemble a turkey's tail from the back, and they also wore a headdress of turkey feathers (Mails 1973:149). The Assiniboine Grass Dance, which featured the rooster tail belt mentioned previously, also incorporated a roach of turkey feathers from the beard of the turkey into their dance cap, as well as a single Golden Eagle feather attached to a bone tube (Van-Stone 1996:11, 54). For the Dog Society headdress of the Hidatsa, turkey or hawk feathers created the fan at the back of the headdress (Mails 1973:167). It is interesting to note, however, that there is only limited archaeological evidence for the use of Wild Turkey feathers along the Missouri trench (Parmalee 1979:211–212). Despite the major amount of work conducted in the Heart and Knife River regions, no turkey remains have been found. This absence raises a question: were turkey feathers a trade item?

The only mention of grouse feathers in regalia comes from the Blackfoot Kisapa Society Dance. The Kisapa society was a social organization of young men. As part of their dance regalia they donned "arm-bands of deer skin and brass, with pendants of grouse and woodpecker feathers" (McClintock 1999:272). It is possible that grouse and prairie-chicken feathers were used in other regalia, especially considering the role of the prairie-chicken in several origin stories as noted in Chapter 4. Since animals associated with origin stories are often incorporated into the regalia of the ceremony that commemorates the story, it is likely that the prairie-chicken was used more in regalia at some point. The decimation of prairie-chicken populations through overhunting and habitat encroachment has negatively affected this bird and its availability for human use.

CHAPTER NINE

Bird Objects in Bundles and Ceremonies

Zedeño (2008a:362) proposes that "few objects are better suited for exploring relationships among people and their surrounding than bundles." A bundle is a complex collection of objects that are kept wrapped together, often in skin or cloth, thereby retaining heightened power by their proximity and association with one another. Bundles are often gifted through a dream or vision by a spirit who has chosen to give some of its power to a person (Hanson 1980; Wissler 1912:103); or a bundle may have its origins in one of the major creation stories for a tribe. The vision explains what should be included in the bundle, associated songs, dances, movements, prayers, as well as instructions for transfer, and other obligations to the bundle and the spirit (Zedeño 2008a:364). Each bundle has a life history, beginning with the dream or vision, and matures in complexity and power over time through its use in ceremonies, maintenance and replenishing of its contents, and transfer from person to person. This power is passed on with the bundle when it is transferred, inherited, or purchased.

Bundles are regarded by the tribes as object-persons, complex objects that are actually alive (*animate*) with a spirit or soul and have the power to influence both people and the environment (Hanson 1980:201; Zedeño 2008a:364). Not only are bundles complex and animate beings, but their power can be transferred from a bundle to a person and vice versa. The ideas of an object's animacy, transfer of power between objects and humans, and transmutation (the ability of humans and objects to take entirely different forms, such as a person transforming into an eagle, a raven becoming a person) are critical concepts in native worldviews in the Missouri River Basin. These ideas begin to illuminate the significance of the bundle system not only in ceremonial life, but also in

governing day-to-day practices such as arrow-making and gardening (Bowers 1992, 2004; Zedeño 2008a:366).

When speaking of the Crow, Lowie remarked, "[l]iterally, almost all medicines are bundles, i.e., wrapped up aggregates of sacred objects" (Lowie 1922b:391). Following Zedeño's treatment of Richert's (1969) typology of bundles (Zedeño 2008a:364–365), however, there are several different kinds of bundles used by the Plains Indians: personal, medicine, and ceremonial, each with an increasing level of complexity in its contents, ritual, transfer practices, and power.

Personal bundles are "biographical" in nature, commemorating significant experiences such as visions, crises, and personal triumphs (Zedeño 2008a:364). As noted earlier, the most familiar example is use of personal medicine objects or "charms" worn by warriors as they went into battle, either to protect them or to imbue some of the power from the object's spirit. Personal bundles often stay with the individual until death or may be transferred to a family member or friend.

Medicine bundles are composed of objects that are used for a specialized purpose, such as horse medicine (Figure 9.1), doctoring, or rain-making. Although these bundles are owned by individuals, they can be used for both individual and community purposes (Hanson 1980:203–204, table 2). One example of a medicine bundle comes from Lowie's account of a Crow policeman who used his medicine from the whirlwind ghost to recover lost property for a couple who had asked for his help (Lowie 1922b:380–381). Medicine bundles are transferred from one individual to another, after which the previous owner relinquishes the power and rights associated with the bundle (Wissler 1912:104).

Ceremonial bundles are the most complex and powerful of all the bundles types. They are associated with ritual

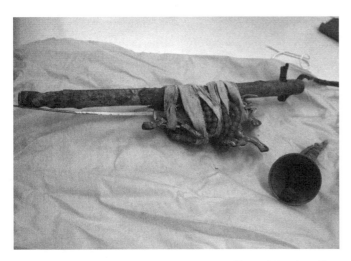

Figure 9.1. Mandan horse medicine bundle with bird quill (2005). Photo by B. Pavao-Zuckerman.

enactments of the creation of the world, a society, or a clan; or they memorialize an important event (Zedeño 2008a:365). Each individual object in the bundle holds associations with people or events that were part of the bundle's creation. In addition, ceremonial bundles may have added complexity through the incorporation of other bundles into the ceremonial bundle or they may be split into separate bundles. Ceremonial bundles represent a record of the collective history of important people and events in cultural memory, as well as a "channeling of cosmic power" through the union of the material and spiritual worlds in an animate bundle (Lokensgard 2010; Zedeño 2008a:366). They are passed down through time according to particular systems of transfer such as—formal ceremonial exchange, inheritance, clan membership, personal merit, and so on. The specific system varies by tribe.

Bird feathers, parts, and skins are among the most pervasive contents in bundles possessed by Missouri River tribes. Bundles are currently in use, although many still are in museum collections. Ethnographic accounts and, rarely, archaeological evidence, support the use history of these complex items. Each tribe has a specific bundle system. Zedeño (2008a:365) argues that bundles are "the embodiment of both the physical world and the rules and regulations of the cosmic, natural, and social orders.".

Bird elements are included in bundles alongside other important materials with powerful spiritual properties, some of which are shared among tribes in the region, others of which are important to a particular tribe. For example, tobacco and red paint are both important in Plains culture, red paint for its healing and protective qualities as well as its ability to bring objects to life, and tobacco as a conduit for interaction between the human and the supernatural (Zedeño 2008a:372, 2009:412). Both items are commonly found in bundles across the Plains alongside bird feathers and other parts. Bird materials may also be found in bundles with items of significance to a particular tribe. For example, corncobs may be found in Arikara bundles, representing Mother Corn, but not in Blackfoot bundles because the Blackfoot do not cultivate corn. The breadth of contexts provides evidence for the widespread importance of birds, but also the varied significance of birds in relation to a particular group's experiences, beliefs, and cultural practices.

Most of the objects mentioned in this section are components of ceremonial bundles used in the major rituals of certain groups, clans, or tribes at large. The following section focuses on some of the ceremonies and associated bundles for which there is the most information, followed by other important ceremonial material culture organized by bird. Bundles sometimes include pieces of regalia such as robes, headdresses, feathers, and other adornments. These were discussed in the preceding chapter in the section devoted to regalia, although some mention of regalia is also included in here, where warranted.

ARIKARA BIRD CASES (OR SACRED BUNDLES)

The Arikara "bird case" is a bundle group containing one of the largest varieties of specifically identified birds. Maximilian witnessed the opening of a bird case and described it in detail:

One of their greatest medicine feasts is that of the bird case, which they have faithfully retained; they esteem this medicine as highly as Christians do the Bible. It is the general rule and law, according to which they govern themselves. This instrument is hung up in the medicine lodge of their villages and accompanies them wherever they go. It consists of a four-cornered case, made of parchment, six or seven feet long, but narrow, strengthened at the top with a piece of wood. It opens at one end, and seven schischikués or gourds are fixed at the top, ornamented with a tuft of horse-hair dyed red. . . . Inside the box there are stuffed birds of all such kinds as they can procure; that is to say, only such species as are here in the summer. Besides these, the box contains a large and very celebrated medicine pipe, which is smoked only on extraordinary occasions and great festivals. If an Arikara has even killed his brother, and then smoked this pipe, all ill-will towards him must be forgotten. With this singular apparatus a ceremony is performed as soon as the seed is sown and the first gourds are ripe. . . . For this they must take down and open the bird case, on which occasion medicine songs are sung and the large pipe is smoked. In the summer-time when the trees are green, they take an evergreen tree, such as a red cedar, peel the trunk, and paint it with blue, red and white rings, and then plant it before the medicine lodge; the case is taken down, and the ceremony performed. This bird case is of special efficacy in promoting the growth of maize and other plants; and he who carries this magic case to a great distance, and with considerable exertion, obtains the highest place in the favor of the Lord of Life. The strongest men among these Indians are said to sometimes carry a whole buffalo, without the head and intestines, to present it as an offering to the bird case. This offering is considered very meritorious; and, when they have made it four times, it is believed that they will never be in want of buffalos" [Thwaites 1906:391–392].

Maximilian's description has been widely cited throughout the regional ethnographic literature (see for example, McGowan 1942:115–117; Parmalee 1977b:219; Will 1934:15), and is supplemented by other accounts from Gilmore (1932) and Will (1934), who described it as a "Sacred Bundle."

Gilmore explains that each Arikara village had their own sacred bundle. Its contents varied from village to village but each had basically the same functions (Gilmore 1932). Each sacred bundle contained a sacred pipe, tobacco, a mussel shell used with the tobacco, ears of corn, a hoe made from buffalo bone and box-elder wood, and five gourd rattles (Figure 9.2; Gilmore 1932:36).[1] Along with these items, the bundle contained numerous bird skins, earning its name as a bird case, as well as the skins of other small mammals and fish.

Gilmore's (1932) accounts of various Arikara Sacred Bundles are truly remarkable. His bird identifications are quite reliable, as not only was he able to handle, examine, and photograph the contents of the bundles, but he also brought Mr. Russell Reid, a trained ornithologist specializing in North Dakota birds, to assist him. Ordinarily, sacred bundles would not be opened in the presence of someone outside the tribe. However, the owner of this particular

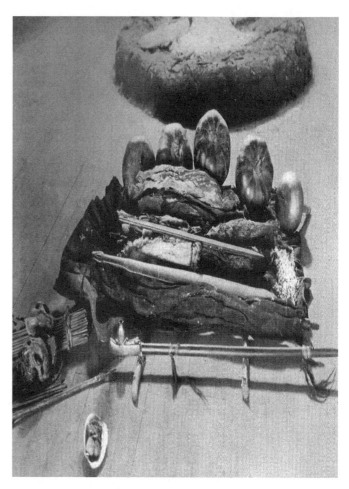

Figure 9.2. Arikara Sacred Bundle. Though there are no obvious feathers in this bundle, there are bird skins just below the gourds (date unknown). Source: Gilmore 1932:Plate III.

bundle had passed on before sharing the knowledge for the ritual associated with that bundle and it had thus been retired. Retired bundles still hold significance for the tribe even though they are no longer actively used for their original ceremonial purpose. Continued bundle maintenance by a custodian with the rights and knowledge to open and renew the bundle is required so "that it might rest in greater security and receive the respect it deserved, and that we might hold in honored and grateful memory those whose lives and thoughts and dreams in the distant, vanished past were bound up in it" (Gilmore 1932:39). Maintenance of this retired bundle afforded Gilmore the opportunity to examine its contents so closely.

The species listed for this bundle included eagle feathers (species unspecified), one Western Grebe (*Aechmorphorus occidentalis*; Plate 7), one small "northern loon" (Common Loon), one Long-eared Owl (*Asio otus*), four Burrowing

---

[1] Maximilian records seven gourds in the bird case, so the number may have varied over time or between bundles.

Owls, one Swainson's Hawk (*Buteo swainsoni*), one "duck hawk" (Peregrine Falcon [*Falco peregrinus*]), one Cooper's Hawk, and one Carolina Parakeet (*Cornuropsis carolinensis*)(Gilmore 1932:40). Carolina Parakeet skins were used by several Plains tribes (Parmalee 1977b:214). Now extinct, the bird typically ranged as far north and west as Kansas and Nebraska, although Maximilian reported the "Carolina parrot" in the vicinity of Fort Clark in North Dakota in the early 1830s (Thwaites 1906:250). Another report of an Arikara sacred bundle recorded the presence of "at least six skins of the extremely rare Carolina Parakeet" (Strong 1933:73, as cited in Parmalee 1977b:214). Plains people would have valued these birds for their rarity, as well as their brightly colored plumage of vivid greens, yellows, and reds. Among the Blackfoot, Little Dog owned a sacred pipe with the body of a parakeet affixed to it (Scriver 1990:263). Thus, the Carolina Parakeet's presence in ceremonial bundles and other items suggests that rare birds also hold significance in native belief systems.

Gilmore examined another sacred bundle from the Arikara village of Hukáwirat that contained a ceremonial meat hook made from a hawk claw (unidentified species), one American Kestrel (sparrowhawk) skin with shell bead eyes, two Swainson's Hawks, one unidentified hawk (*nikritawikrisu* in the Arikara language*)*, another larger unidentified hawk species, four feathers from a Snow Goose (*Chen caerulescens*; Plate 4), a small species of owl, one Burrowing Owl, and possibly a kingfisher (1932:44–45).[1] In addition to the gourd rattles, there were 34 willow sticks, a large catlinite pipe, 4 ears of corn, a mussel shell, native tobacco (*Nicotiana quadrivalvis*) and sweetgrass (*Hierochloe odorata*), another smaller bundle of calico inside containing corn, ground vegetable powder, a piece of Indian hemp cord (*Apocynum cannabinum*), mother-wood, a feather from each species of bird in the larger bundle, and an eagle bone whistle and down feather (Gilmore 1932:44–45). The 34 willow sticks were laid out during the Sacred Bundle ceremony to illustrate the origins of the world and the Arikara people. The willow sticks that pointed to the southwest were to remind them of Thunder and the water of life, including waterbirds (Gilmore 1930:105, 116–177).

The willow sticks that pointed northwestward represented the wind as the "breath of life and all the powers of the air," including birds and insects (Will 1934:15).

The objects in bird cases provide a material record of the Arikara origin story (Gilmore 1930:116–117). When the Arikara emerged from underground, led by Mother Corn, they faced many obstacles on their journey westward. At each obstacle, Mother Corn called to the gods for help and a bird came to help the Arikara continue. A kingfisher helped them cross a chasm by collapsing the high walls and forming a bridge. An owl blew down the trees to create a path through the forest, and a loon flew over a lake to a pathway (Dorsey 1904:12–25). Moreover, at each obstacle, some Arikara chose to stay behind and become that kind of bird (Dorsey 1904:14–15), making the birds their kin.

ASSINIBOINE SUN DANCE

The Sun Dance is well known throughout the Plains and each tribe has their own traditions and meanings attached to the ceremony. The eagle plays a central role in the Assiniboine Sun Dance, appearing in imagery on the lodge poles and in ceremonial objects used during the four-day observance held for the "spirits of the Thunder Bird in all their forms" (Rodnick 1938:48). The Sun Dance was practiced annually for a host of reasons involving the general health and prosperity of the tribe, such as successful hunting and plentiful food, guarding against illness, safety of warriors, and generally good morale for the community (Rodnick 1938:48).

The symbolism of the center pole for the Sun Dance Lodge, made from a cottonwood tree, demonstrates the principal importance of the Thunderbird to the ceremony. The selection and chopping down of the tree were sacred activities, accompanied by prayers and smudging (Lowie 1909:60). Once the tree was dragged to the selected site of the Sun Dance Lodge, it was carved with the images of the Sun and an eagle representing the Thunderbird near the top, and a buffalo head near the base (Figure 9.3; Lowie 1909:61). The branches of the tree were assembled at the top of the pole and tied with skin to represent an eagle nest. Here an offering would be made to *Wakan'-taṅga* (the "Great Mystery") by hanging red flannel from a fork in the branches at the top of the pole (Lowie 1909:61–62). A similar offering was made toward the Thunderbird as well. In a letter describing the religion of the Assiniboine, Jesuit missionary Pierre-Jean De Smet commented that "Thunder, next to the sun, is their Great Wah-kon" (Chittenden and Richardson 1905:936).

He later continues:

---

[1] The Arikara believe that the Burrowing Owl used to live in the forest with other owls, but that over time its habits changed and it moved onto the prairie to live in the abandoned burrows of prairie dogs and other small mammals. The owl skin contained in the bundle is wrapped with a twig of "mother-wood," an unidentified species the Arikara call *nakis-atina*, to demonstrate the Burrowing Owl's ties to the forest (Gilmore 1932:43–45). Gilmore reports that the bird resembled a kingfisher, but was larger in size that the Belted Kingfisher, which is the most common species in the region.

Figure 9.3. Gus Hegelson places a tobacco offering at an Assiniboine Sun Dance lodge. The center pole is carved with a Thunderbird and lightning bolts. Photograph taken by Lisa Hornstein©, Native News Project, Fort Belknap Reservation, Summer 2004. Source: http://nativenews.jour.umt.edu/archives/nativenews2004/feature_articles/Fort_Belknap/Fort_Belknap.html.

The Sioux, or Dakotas, of whom the Assiniboins are a branch, pretend that thunder is an enormous bird, and that the muffled sound of the distant thunder is caused by a countless number of young birds! The great bird, they say, gives the first sound, and the young ones repeat it: this is the cause of the reverberations. The Sioux declare that the young thunders do all the mischief, like giddy youth, who will not listen to good advice; but the old thunder, or big bird, is wise and excellent, he never kills or injures anyone! [Chittenden and Richardson 1905:945].

There is little information discernible from the early descriptions of ceremonial objects or other material culture associated with the Assiniboine Sun Dance. De Smet mentions that the Assiniboine prepared regalia and other ceremonial articles throughout the winter in preparation for the Sun Dance, "from the embroidered legging and moccasin to the eagle-plumed head-piece" (Chittenden and Richardson 1905:939). Lowie also mentions the use of eagle-bone whistles and eagle feather fans during the ceremony (Lowie 1909:59, 61). Eagle feather fans were specifically used for fanning the fire in the preliminary lodge that was used for preparations before the people went out to select the Center Pole for the Sun Dance Lodge. They believed that if the people blew on the fire rather than using a fan, it would cause a windy day (Lowie 1909:59). The presence of other birds in ceremonial objects of the Sun Dance is not well documented.

BLACKFOOT THUNDER MEDICINE PIPE

There are several Blackfoot accounts of the acquisition of the Medicine Pipe and its ceremony. Common throughout each story, however, is a battle of the powers of Raven and Thunder, and the gift of Medicine Pipe and bundle by Thunder after his defeat (Grinnell 1913: 53–59, 1920:113–116; McClintock 1999:253; Zedeño 2008a:369–370).

"Now," said the Thunder, "you know me. I am of great power. I live here in the summer, but when winter comes, I go far south. I go south with the birds. Here

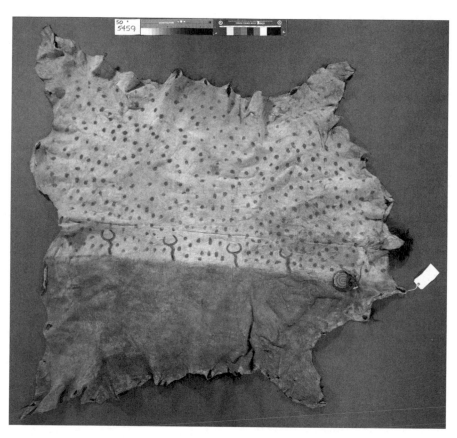

Figure 9.4. Blackfoot Medicine Pipe owner's robe, decorated with the hailstone motif and Thunderbird tracks, c. 1904. AMNH 50/5459.

is my pipe. It is medicine. Take it, and keep it. Now, when I first come in the spring, you shall fill and light this pipe, and you shall pray to me, you and the people. For I bring the rain which makes the berries large and ripe. I bring the rain which makes all things grow, and for this you shall pray to me, you and all the people." Thus the people got the first medicine pipe. It was long ago [Grinnell 1920:115–116].

Similar to the Arikara sacred bundles, there are many Thunder Pipe bundles; however, there is a hierarchy of importance in the Thunder Pipe bundle system (Zedeño 2008a:368). Running Wolf received from his father, a Thunder Medicine Pipe priest, the "Long Time Pipe" (McClintock 1999:427). This pipe retains a special status for the power stemming from its origin as a direct gift from the Thunder, as well as the power gained from its transfer from chief to chief over generations. As with other important ceremonial bundles, transfer of the bundle over time creates a life history, and through this trajectory the bundle accumulates additional power as the relationships among objects within the bundle become more complex and intertwined. Zedeño (2008:368) explains that "the power of bundles lies in the cumulative sentience of each and every object within it" as well as in the web of relationships among objects, which together become more than the sum of parts.

The Thunder Medicine Pipe bundle contains the skins of many animals and birds wrapped in the skin of a black or grizzly bear, which the bear offered as part of the bundle in the origin story. Scriver (1990) lists animals such as "raven, crow, magpie, golden eagle, loon, crane, muskrat, prairie dog, squirrel, mink, and elk," each with their own song and ceremonial role in the Medicine Pipe ceremony. The owner of the Thunder Medicine Pipe bundle keeps a robe that is decorated with red paint on one half and the distinctive Blackfoot design of Thunderbird tracks and hailstones on the other half (Figure 9.4). This design has been passed down through the transfer of the bundle and has been used on other ceremonial objects and even tattoo designs for bundle holders, emphasizing the centrality of the Thunderbird to the bundle and ceremony over time (Zedeño 2008a:372–373).

The bundle is opened each Spring, with the roll of the first thunder. Just last year a Blackfoot Thunder Pipe owner related that "my pipe opening ceremony went really excellent. As soon as AP brought the pipe outside the lodge we heard the crack of Thunder. He was letting us know that he was here watching the ceremony." Many birds are called on during the transfer ceremony for the Thunder Medicine Pipe bundle, including the owl, grouse, Thunderbird, swan, crane, ducks and geese, and "good rusher" (McClintock 1999:265). The owl plays an important role in the incorporation of new members into the Pipe Society, vital to the element of surprise in capturing a new member (McClintock 1999:254). The owl's favorite food, the "Siksocasim-root," was offered during the ceremony in order to please the owl and ensure its assistance in acquiring a new member. James White Calf, a consultant of Claude Schaeffer during his work with the Blackfoot during the mid-twentieth century, explained that Medicine Pipe owners turn into owls when they die (Hungry-Wolf 2006:139). James White Calf was visited by a deceased Medicine Pipe owner while camping on the Marias River with Fish Child. An owl approached them during the night, and they offered him the pipe to smoke by filling it with tobacco, lightning it, and setting it outside their tent on the ground. Afterwards, the Owl told them he was going away. When they retrieved the pipe, the tobacco had been burned out. According to James, "That is why the Medicine Pipe owners started to include an owl skin in their bundles to hold tobacco thereafter" (Hungry-Wolf 2006:139).

> It's a significant thing [the owl] because a long time ago the women would not dance with the medicine pipe but they'd dance with the owl. There are songs, you know, that belong to the owls and the medicine pipe. Of course all the birds all have songs too that belong to the Beaver Bundle. But the owl, you won't find him in the Beaver Bundle but you'll find him in the Medicine Pipe Bundles. There's even ways of signifying that you are a Medicine Pipe [person] because of how you grab that Medicine Pipe, you grab it like an owl.
> —Blackfoot consultant

The grouse is honored in the Medicine Pipe ceremony because it had "given its power to the Pipe" (McClintock 1999:259). Drums, decorated with symbols of the sun, moon, stars, birds, and animals, were played in imitation of the territorial "drumming" display of the male grouse. The Thunderbird, swan, and crane were all recognized in the ceremony for transferring the Medicine Pipe to a new owner through the imitation of their flight or call during each birds' respective song and dance (McClintock 1999:265). There were no dances for waterfowl, but songs were sung for the ducks and geese. Later in the ceremony, there were four Horse songs, seven Owl songs, seven Buffalo songs, and seven songs for a diving water bird called the "Good Rusher," named for its ability to run across the surface of the water (McClintock 1999:265).[1] The Good Rusher is regarded as very powerful, with the ability to drown people by dragging them beneath the water, suggesting that this bird is an adept diver. Significantly, the owl and good rusher receive similar recognition in the Pipe transfer ceremony as the horse and buffalo, two animals that are much more widely recognized as central to the Plains tradition. The Medicine Pipe bundle transfer ceremony is a tradition that continues today, particularly after the repatriation of sacred pipe bundles (Conaty 2015). The ceremony recognizes the significant responsibilities and respect the community has toward the bundle as a powerful, living force.

BLACKFOOT BEAVER BUNDLE

The Beaver Bundle has several varying stories relating to its origin, but birds are always present in them. The role of birds is greatest in the Piikani and Blood versions of the story recorded by Wissler and Duvall (1908:74–76). In bundle origins, Beaver instructs a man to invite the Sun and Moon to his lodge if he wants the medicine of the Beaver bundle. The man pays the Sun in eagle-tail feathers, hawk-tail feathers, and black-fox hides in return for receiving the ritual songs for the bundle. The Sun then gives the man instructions to use the bundle in the spring during the planting of tobacco, along with the performance of a tobacco-seed dance to ensure a successful tobacco crop. In the Blood variation, a woman was impregnated by a Beaver and gave birth to a baby beaver, which she kept in a bowl of water at the head of her bed. As a reward for the husband's kindness to the beaver baby, the Beaver decided

---

[1] The Good Rusher could refer to any number of diving water birds, including the cormorant, grebe, or loon. Schaeffer's (1950:39) translation of the Blackfoot word for the loon (*Matsiisaipi*) is "fine charger," a synonymous phrase that suggests it is likely referencing the same bird. However, Schaeffer also comments that some informants used the term to refer to mergansers, so its exact reference remains ambiguous. One of our Blackfoot consultants identified the loon as the "fine charger," consistent with Schaeffer's notes.

to give the man his Beaver medicine-songs in exchange for bird and other animal skins.

> After a while the Beaver began to sing a song in which he asked for skin of a certain bird. When he had finished, the man arose and gave the bird-skin to him. Then the Beaver sang another song in which he asked for the skin of another bird, which was given to him. Thus he went on until he secured all the skins in the man's lodge. In this way the man learned all the songs that belonged to the beaver-medicine and also the skins of the animals to which the songs belonged. After this the man got together all the different kinds of bird and animal skins taken by the Beaver, made them up into a bundle, and kept the beaver-medicine [Wissler and Duvall 1908:76].

McClintock's (1999) account of the Beaver Medicine ceremony provides details of the ceremony and bundle, as well as another story for the origins of Beaver Medicine, in this case given to a man who was deserted on an island by his brother, and invited to live with the beavers there and learn about their medicine.

The Beaver Bundle is important to tobacco planting rites, and this link has an origin story that explains the relationship between the bundle and tobacco-planting, with an emphasis on the role of birds (Wissler and Duvall 1908:79–80). A Beaver Bundle owner went off on a journey, asking the other bundle owners to wait for him to return before they planted tobacco. The men did not listen, and when the bundle owner returned from his journey he saw that they had already planted their tobacco seed. He was grieving alone out on the prairie when he was met by an old man who called together many birds and animals to help the bundle owner plant tobacco. The animals offered their dung to fertilize the plants; however, they had no tobacco seeds to plant. Here, birds become integral to the acquisition of tobacco seeds from the Sun:

> As they had no tobacco-seed, three birds volunteered to go to the sun for a supply. So they started off to the sky, and when the first cloud was reached, one of them gave out; but the cloud turned him yellow. When the next cloud was reached, another gave out; but he became red. The other bird went on alone until he finally reached the sun and became black. This one brought down the seed. Now all the animals assembled, and proceeded to plant the new tobacco-seed. They sang many songs, and performed all the parts of

the ceremony. When the seed was finally planted, they fenced in the plot with rocks and sticks, after which they all went away [Wissler and Duvall 1908:79–80].

At the end of the season, the tobacco planted by the other men had been trampled by buffalo, while the seeds planted for the man by the birds and animals had flourished. He later invited all the other Beaver Bundle holders to his lodge, where they transferred their bundles to him and the tobacco planting rites have been a part of the Beaver Bundle ever since.

According to Wissler's (1912) account, there are a variety of birds kept in the Beaver Bundle, of which there is a "primary bundle" as well as other bundles among the divisions of the Blackfoot Confederacy. The specific contents vary from bundle to bundle. He mentions that "skins of loon, yellow-necked blackbird,[1] raven, blackbirds, woodpeckers, sparrows, crow, ducks, and several birds we were unable to identify" were important inclusions (Wissler 1912:169). One Blackfoot consultant described the loon as one of the principal animals in the Beaver Bundle: "The loon, 'fine charger,' is very significant in the Beaver Bundle. Kind of like a chief. The loon and the otter are used to bring rain by the Beaver Bundle people. It's rain medicine." Another Blackfoot consultant told that religious leaders gave her and her husband the charge of reconstituting a Beaver Bundle. In the transfer, they only received the bundle pipe, which had been repatriated to the tribe, and the instructions about how to go about reconstituting the bundle's essential and complementary components. The loon, the elders told them, was indispensable.

Other principal contents included red paint, beaver skins, buffalo parts, weasel, white gopher, prairie dog, and immature antelope, deer, mountain goats, sheep, domestic dog, and Iniskim fossils. One example of a particular Peigan Beaver Bundle included a loon, eagle and raven feathers, "white prairie-chicken, yellow-necked blackbird, mud hens or hell divers, hawk, white swan, ten sparrows, [and] snow bird," while another, very old bundle contained a rail, two wall creepers,[2] grouse, a nighthawk, and magpie feathers in addition to the birds already mentioned (Wissler 1912:170). The Beaver Bundle also included a

---

[1] Commonly known today as the Yellow-headed Blackbird (*Xanthocephalus xanthocephalus*; Plate 45; Sterry and Small 2009:374).

[2] Possibly the Brown Creeper (*Certhia americana*). Wissler remarks that one of the creepers is black, and Sterry and Small (2009) confirm that the subspecies ranging in western North America is darker in color, smaller, and longer-billed than other varieties.

Water Pipe and sacred digging stick, and sometimes a bundle of counting sticks used as a calendar. Schaeffer discusses the counting sticks in his unpublished notes on the Beaver Bundle (Schaeffer and Schaeffer 1934: GM, Blackfeet M-1100-129). The counting sticks have representations of bird or animal heads on one flattened end of each stick, each holding a particular meaning for the month. Schaeffer adds that the "wisest being for weather is the loon, whose stick is kept in the middle of the sticks and not turned,"[1] and that not all bundle owners are able to use the counting sticks.

The Beaver Medicine ceremony complements both the origin stories recorded by Wissler and Duvall (1908) and McClintock (1999). Sweetgrass is burned at the beginning of the ceremony, mirroring the Sun's burning of sweetgrass before he began to give the Beaver songs to the man. Many birds are mentioned in the ceremony, including the two songs for the crow, a wild goose song, a song for the Mallard and the inclusion of a Mallard head in the bundle, and the songs of a Red-headed Woodpecker and the presence of its skin in the bundle. The Red-headed Woodpecker has three songs for the Beaver Bundle, because "in the beginning, when the birds gave their songs to the Beaver Medicine, the woodpecker gave three" (McClintock 1999:98, 111). Bird songs are interspersed with songs and dances recognizing other important animals for the Beaver Bundle, including the beaver, elk, moose, buffalo, lynx, badger, and otter. The ceremony closes with the dog dance. Today, ceremonies of the Beaver Bundle are gathering strength with every transfer, and many young couples in the Blackfeet Indian Reservation and other reserves in Canada are taking vows to become Beaver People.

CROW SUN DANCE

The use of eagle materials and imagery in the Crow Sun Dance ceremony provides striking evidence of the importance of the eagle in Crow cosmology and ceremonial practices. The eagle's connection with the Sun is repeatedly referred to in the literature, referencing the eagle as a "messenger" between humans and the Sun (Lowie 1919, 1922b; Voget 1995). Therefore, the eagle's paramount role in the Sun Dance is as a conduit of communication and power between the participants and the spirit world, particularly the Sun. There is widespread use of eagle parts,

although other birds also contribute significantly to the Sun Dance Ceremony through their inclusion as ceremonial objects.

Medicine Dolls

Lowie's 1915 account of the Crow Sun Dance details the need for a medicine doll in order for the person who pledged the Sun Dance to achieve the "desired vision" for the ceremony. The origin story of the medicine doll indicates that it came to the Crow in the guise of a screech owl who emerged in its doll form and took its place inside the Sun Lodge through the magic of songs sung by the Moon and her seven men helpers (Lowie 1915:14). The pledger or "whistler" of the Sun Dance had to entreat a man who was a doll owner to preside over the ceremony, taking on a ceremonial relationship of father and son. Medicine dolls were received through a vision that instructed the man on how to make the medicine doll, and each doll was kept in its own bundle (Figure 9.5). The medicine doll was often adorned on its head with plumes of feathers; one doll viewed and photographed by Lowie (1915:16) was made with so many feathers that "the entire figure was almost covered out of sight with a profusion of owl feathers." A feather from the medicine doll might be taken into battle and worn by a warrior for good luck, but its main purpose was associated with the Sun Dance.

The origins of the first medicine doll demonstrate a close relationship with the screech-owl and the reason for the presence of owl feathers on many medicine dolls. *Andícicòpc* first discovered the medicine doll with the help of a "little bird" who came to him as he was traveling near Billings, Montana, and told him to look toward Mount *I'ExuxpEc*, where he saw seven men and one woman drumming and singing. The woman, who was the Moon, was holding a medicine doll and they taught him the songs associated with it. Later the group appeared at his bed with the Medicine Doll Bundle. As they sang, the doll came to life, revealing itself more and more with each song.

> After the fourth song, the woman stepped forward and then back again. The doll came out completely in the guise of a screech-owl, and sat down on the moon's hand. The boy was at this time lying straight on his back. The screech-owl flew about, and then perched on Andícicòpc's breast. Suddenly one of the men loaded and cocked a breechloader, then he stepped toward the boy and sang a song. The woman said to the screech-owl, "Now, little screech-owl, this man is going to shoot you, you must make your medicine." It stood up

---

Figure 9.5. Crow Sun Dance doll decorated with owl feathers (date unknown). AMNH 50.1/ 4011A (also in Lowie 1915b:16).

on its feet and began to flap its wings. The man drew closer, and shot at the owl, which entered his breast and began to hoot inside. Andícicòpc looked towards the northeast. In the valley he saw a sun dance lodge [Lowie 1915b:14].

Sun Dance Lodge Poles

The main poles of the Crow Sun Dance lodge could also be considered "ceremonial objects" that intimately link eagles with the ceremony. According to Lowie's 1915 account, there is no "center pole" as is often described in the Sun Dance of other Plains tribes, although the first tree to be cut was a cottonwood and would be left in the forest where it fell (Lowie 1915b:32; Lowie 1954:178–179). In Yellowtail's contemporary account of the Crow Sun Dance, he does

reference a center pole that receives its own special prayer before it is cut (Fitzgerald 1991:145–146). The center pole is worshiped with a pipe and eagle feathers, which are smudged and rubbed on the tree during the prayer (see Figure 6.1).

Other poles were cut from cottonwood or pine trees depending on the vision, with three or four poles serving as the main poles for the Sun Dance lodge, according to Lowie's early account. These poles, wrapped in buffalo hides, willow sticks, and ground cedar (*Lycopodium complanatum*), represent an eagle's nest. During the construction of the Sun Dance lodge, a man with eagle medicine would climb up into the symbolic nest; Lowie (1915b) refers to this person as the "bird man." The bird man wore a robe fastened with an eagle wing-feather and brought an eagle bone whistle and eagle feather fans up into the eyrie, where he sat imitating an eagle by flapping the feather fans and whistling like an eagle (Lowie 1915b:37–39). The bird man remained in the nest at the top of the lodge while the remaining poles were raised, after which he descended his perch and announced a vision of a slain enemy to the whistler.

During the Sun Dance, eagle medicine objects were also important to the No-Fire Dance (*birǎ'retarisùa*), when war leaders and their followers danced to attain a vision for the whistler (Lowie 1915b:39–41). In one account of a war captain's vision, a widow for whom the Sun Dance was being held, requested that a war captain with eagle medicine lead his war party in the No-Fire Dance. He used an eagle-bone whistler and his eagle medicine, a bald-eagle head, along with a bald eagle song, in order to obtain a vision of slain enemies, which was fulfilled four days later.

Contemporary accounts of the Crow Sun Dance from Fitzgerald (1991) and Voget (1995) emphasize the role of medicine men in healing the sick during the ceremony. A Crow consultant told the authors that people often affix pieces of their own flesh as vows to the Sun, to receive health and other blessings. The center or main poles of the Sun Dance lodge are a source of power from the Creator, which the medicine man can use to draw sickness out of a person with medicine feathers (Voget 1995:199).

The center pole is used by the whistler and the medicine man in "sending prayers and receiving messages from the Sun," a bond strengthened by the symbolic eagle's nest on top (Voget 1995:199). Yellowtail recalled a Sun Dance held by John Trehero, which exhibited the immense power that could be transmitted by the center pole of the Sun Dance lodge through eagle medicine feathers (Fitzgerald 1991:67). A non-Indian woman in attendance at the Sun

Dance repeatedly voiced her doubts about the healing power of the Sun Dance ceremony and the medicine man. Hearing the criticism, Trehero exhibited the power held in his eagle feather fan:

> He took the little feather fan and went up to the center and made medicine where there is great power, on the center of the tree. He put more power into the little feather fan by putting it against the center tree, at the base of it. That woman sat there and watched and kept criticizing. She was some thirty feet away from where John was, at the doorway among the people. Finally, in the direction of where she was standing, he pointed right at her, like that. He didn't toss it, he just held the feather fan. He didn't want to really hurt her, so he didn't let it go. His power from that far away struck that woman as if she was shot by a gun. She dropped down like a shot deer among the crowd [Fitzgerald 1991:67].

In addition to eagle parts, Yellowtail also used a stuffed eagle in the ceremony. He acquired the eagle from his brother's ranch, believing that "the eagle wanted to give himself to [Yellowtail] for the Sun Dance . . . 'cause this great bird came down from the sky into a coyote trap" and was found in perfect condition (Fitzgerald 1991:153). Yellowtail emphasized the importance of using buffalo and eagle parts in the Sun Dance ceremony in order to "bring blessings" to the tribe (Fitzgerald 1991:154). Yet, eagle parts and especially whole eagles are difficult to acquire because of the federal regulations (Bald Eagle Protection Act, 16 U.S.C. 668–668d, 54 Stat. 250; Presidential Memorandum 59 F.R. 22953).

CROW TOBACCO CEREMONY

It is believed that the survival of the Crow as a people is intertwined with the continuation of Tobacco planting traditions (Denig 1980:59). Tobacco is regarded as the "distinctive medicine" of the Crow, analogous to the Medicine Pipe's importance among the Hidatsa (Denig 1980:62; Lowie 1919:177). Among the Crow, planting tobacco is not a right, but a ceremonial privilege typically relayed through a vision or adoption by a member of the Tobacco Society (Lowie 1919:112). Tobacco was originally given to the Crow by a star. Chapters of the Tobacco Society were formed as people received visions related to tobacco.

Some accounts of the origins of tobacco for the Crow are linked with the Crow creation story. The Sun, associated with creator Old-Man-Coyote, gave the Crow instructions for how to plant tobacco soon after the world was created (Lowie 1919:188). A boy was adopted by the Sun and taught the correct practices for planting tobacco, including how and when to plant, songs to sing, and tobacco dances. When they prepared to enter the tobacco lodge, a song was sung to represent the geese who return in the spring, and the man said, "All the birds coming back have a leader. We'll have one too." His wife was given a pipe and led the tobacco processional into the lodge, taking on the role of *akbasáande* or "the one who goes first" (Lowie 1919:188). Women play an important role in the tobacco ceremony for planting, leading the procession to the planting site and carrying the medicine objects (Lowie 1919:163).

Songs associated with the Tobacco ceremony link elements from everything on earth that holds any kind of supernatural power including "elk, antelope, mountainsheep, black-tailed deer, beaver, otter, weasel, wolverine, wolf, coyote, squirrel, eagle, crane, swan, bear, buffalo, horse, eggs, sun, earth, summer, winter, say, night, thunderbird, God, . . . moon, dawn, Morningstar, evening-star, dipper, day-start, and medicine rock" (Lowie 1919:190). Lowie recorded over 30 chapters of the Crow Tobacco Society. There are numerous chapters named for birds: Duck (*míaxākamciɛ́*), Eagle (*nakā'kamciɛ́*), Egg (*í'g·amicɛ́*), Blackbird (*baxíramciɛ́*), Prairie-chicken (*tsítsgè*), White Bird (*nakā'ksiamicɛ́*), one named for an unidentified species of hawk (*ba+ipxaxɛambicɛ́*), Meadowlark (*ma+ūwat-ciramicɛ́*), and Crane, for which no translation was given (Lowie 1919:115). It is unlikely that all of these chapters were functioning on a local level at any one time, with chapters splitting or amalgamating over time. Yet, a consultant who is also a leader in the Sacred Tobacco Society identified these birds as culturally significant now.

While the elaborate ceremonial bundle systems of the Hidatsa and Blackfoot are not as common among the Crow, each chapter of the Tobacco Society has associated bundles or a set of objects for Tobacco ceremonies, as well as individual medicine objects (Lowie 1919). These objects are usually specific to the original visionary's experiences and instructions for the formation of a new chapter, although more items might be added over time.

For example, the Egg Chapter's bundle incorporated a blanket that was dreamed by a poor Crow man and the design purchased by a Crow couple and worn by the wife, a member of the Egg Chapter (Lowie 1919:126). The blanket depicted an eagle in the clouds, understood to represent thunder, and was decorated with several different kinds of feathers. The couple kept the blanket in a bundle, and allowed the visionary to foretell good luck for the couple.

The Blackbird Chapter was founded by Breath, whose father had blackbird medicine and passed it on to him (Lowie 1919:128). He was already a member of the Tobacco Society, but formed a new chapter based on the blackbird medicine. He distributed stuffed blackbirds to those he adopted into his chapter. Members would wear black blankets with stuffed blackbirds attached to them when they danced.

The Eagle Chapter was founded by Sore-tail after he fasted near the Wolf Mountains and received a vision of a special lodge for a Tobacco dance from an Eagle, who he recognized as the Sun's messenger (Lowie 1919:129–130). Sore-tail "hung up an eagle on a shield" in the lodge, and gave the men red-dyed eagle tail-feathers to wear on the back of their heads, and eagle plumes ornamented with blue beads for their left shoulders (Lowie 1919:130). Old men wore special eagle robes with an eagle plume attached to the back. Women wore the smaller wing feathers and carried eagle wings powdered with red paint while they danced. Some of the women wore red dresses with beaded eagles on their shoulder blades. Eagle feather fans were "sacred and were to be moved in different ways during the dance" (Lowie 1919:132). Freely translated, the part of the Eagle Chapter's song calls,

I shall carry my medicines on my back.
All of you, you are poor. Look at me!
I have stayed over the hills, I came from over the hills.
I am an eagle, and I shall walk toward the Wolf Mountains [Lowie 1919:130].

The song makes reference to the eagle medicine in the form of the eagle plumes and beaded eagles on the regalia of the Eagle Chapter dancers. The eagle medicine brought good luck for acquiring horses. The eagle was further invoked in one version of the origin of the Tobacco Society, where the visionary imitated an eagle by singing "his bird song, flapping his eagle wings at the same time" (Lowie 1919:185). Flapping his "wings," he fanned a fire, which spread across the grass to clear a garden for planting the tobacco.

In the story of the origin of the Prairie-Chicken Chapter told by Crane-bear, a Crow man named Blanket-owner was refused admission to the Tobacco Society and went out wandering on the Rosebud. He received a vision of prairie-chickens giving a Tobacco dance with a yellow-painted drum. He made a drum and introduced it into the Tobacco ceremonies of the Prairie-chicken Chapter just like the one he saw in the vision. Crane-bear

continued telling the origin of another Tobacco chapter, the White Birds, in which a young man slept in a tobacco garden and received a vision from a white bird. He began adopting members, and asked one of his adopted members to get the head of the white bird, which has since been passed down through the chapter over time.

In recent times, the Tobacco Society ceremonialism has coexisted with Christian practices. Agnes Yellowtail Deernose told Voget (1995) that her mother continued to participate in the Tobacco Society even after she became a Baptist, as did she for many years. After getting married, Agnes "gave up the Tobacco worship because [her husband] didn't want medicine bags between him and God" (Voget 1995:110). She continued, "The way I look at it now, I don't believe in the old birds that they used in the medicine." Her sister's family continued to stay active in both the Tobacco Society and the Baptist church (Voget 1995:110). This statement is revealing in two ways. First, Agnes' comment about the "old birds" suggests the central role of birds in the Tobacco Society. Second, many Crow believe that practicing their traditional ceremonies are not mutually exclusive to participation in Christian churches, and one can chose to participate in either or both. Not all traditionalists feel that way, though, and three of our elder Crow consultants who are leaders in the Tobacco Society did not approve of Christian practices at all.

In a recent online exhibit entitled "Crow Expressions" on the website for the Western Heritage Center [WHC] (www.ywhc.org) located in Billings, Montana, 12 members of the Crow Tribe gave their perspectives on what it means to be a Crow Indian. Many commented on the prominent role of tobacco in Crow identity. One tribal member observed that "to be Crow Indian is to be connected to the land. This place, where the sacred tobacco seed grew, must continue" (WHC 2008). The Tobacco Society is carried on through various contemporary chapters.

CROW MEDICINE PIPE (SACRED PIPE)

The Crow Medicine Pipe is also closely tied to birds. Their feathers are used on the pipe and their actions imitated in the ceremony. The Crow recognize the Medicine Pipe as being borrowed from the Hidatsa (Lowie 1924:339). The origin story told by Pretty-Horse reflects this and also explains the addition of many bird elements to the Medicine Pipe as requested by the birds themselves. In this account, a group of Hidatsa were returning home from a horse raid when the Dakota overtook them and killed all but one. This Hidatsa man continued his journey home and each night he heard drumming, becoming louder and louder as he travelled.

> He was traveling homeward. During the eighth night he did not sleep at all. He saw people. A pipe about three feet long was lying with its head up. The stem was wrapped with white and blue beads; at intervals bunches of owl feathers were tied along the stem. Duck said, "I want to be in it too." So they stuck the mouthpiece into a duck's neck. Horse wanted to be in it, so he gave the hair from his tail to be tied about the middle of the stem. A person wanted to be in it, so a lock of his hair was taken and tied on. Thunderbird said, "I don't see how I can get into it; I'll make them dance." So he became the drummer, and his thunder is the drumming. Soke wanted to be in it, so he gave his tail for the rattles. Dog said, "I don't care for anything, I have my nose poked into everything. If I am in it, the dance will not be very serious." Even now when the dance is performed, a dog always comes there. The Hidatsa got home. This is how the Pipe dance started [Lowie 1924:340].

A description of another Medicine Pipe by Bear-gets-up included a Red-headed Woodpecker skin tied to the mouthpiece of the pipe, along with red or blue dyed horse-hair. Eagle tail feathers were also attached to the Pipe, in a "dropping fan fashion" (Lowie 1924:348). Eagle feathers from the Pipe were worn by the adoptee during the adoption ceremony, when the Pipe owner ceremonially imitates the feeding of young eagles by moving a piece of meat in a circle during a song.

In the past there were multiple Medicine Pipes owned by those people who were adopted and given a pipe. The Medicine Pipes were kept in bundles along with sweet-grass (Lowie 1924:341), a corncob, wildcat skin, and a plume. Ownership of the Medicine Pipe is an honor but also a burden. Lowie commented that, "most people were afraid to own a pipe for fear of breaking some rule connected with its use." Pipe-owners did not constitute a society, but they did form a community, with shared feasting (Lowie 1924:344). An account by the Crow owner of the last functioning Medicine Pipe, referred to as a "Sacred Pipe," gives an explanation of the significance and decoration of the pipe, drawing much continuity between Lowie's earlier descriptions (McCleary 1997:92–94).

The Crow elder Albert from the Wyola District of the Crow Indian Reservation, the holder of the last Sacred Pipe, also explained in more depth some of the meaning behind the bird materials adorning the Pipe. The owl feathers come from the Great Horned Owl, representing "his power, his good messages" (McCleary 1997:93), while the elder described the eagle plumes as representing "the dog, protector of man." Seven eagle feathers hanging from the pipe stem like a fan represent the Seven Bulls, associated with the origin of the Crow sweat lodge and the Medicine Pipe (Fitzgerald 1991; McCleary 1997:81, 92–94). The Seven Bulls raised a special human boy, who healed them after a battle through the eagle medicine of a sweat lodge ceremony (Fitzgerald 1991:113). To remind the Crow of the significance of purification before and after important tasks associated with sweat lodge ceremonies, the seven bulls decided to become seven stars and formed the constellation of the Big Dipper, the boy joining them and becoming the North Star.

Albert explained the significance of the Mallard head on the pipe stem relating it back to the Crow origin story, where a Mallard dove deep underwater to retrieve mud so that Iichíhkbaalee (First Maker) could make dry land (McCleary 1997:93). Albert specifically drew attention to the blue part of the Mallard's plumage on its head as an important feature; the pipe stem is similarly decorated with blue beads.[1]

MANDAN AND HIDATSA BIG BIRD CEREMONY

The Big Bird rites of the Mandan and Hidatsa, as their name suggests, focus on the movements and supernatural powers of big birds including eagles, hawks, ravens, and crows (Bowers 1992:363, 2004:260). According to oral traditions, the Thunderbird was in some instances considered a relative of these birds and in others was referred to as any one of these groups. Big Bird ceremonies are held in the spring in anticipation of the big birds' migration through the region en route to their summer nesting grounds along streams, in the badlands, and in the Rocky Mountains. Bowers (1992:363) explains that although

---

[1] This plumage is distinctive for identifying Mallards, and the bird is unlikely to be confused with other ducks with any blue coloring on the head (for example, the Northern Shoveler (*Anas clypeata*) or the Red-Breasted Merganser (*Mergus serrator*) because of their noticeable difference in build. This is noted because the Mallard is not typically a diving bird, preferring to "dabble" at the surface by sticking its head underwater to feed in the shallows (Sterry and Small 2009:30; CLO 2011). The Cornell Lab of Ornithology does assert that the Mallard "makes occasional dives into deeper water," although it is not known for that behavior in comparison to other "diving duck" species (CLO 2011).

"there is a close psychological association between Big Bird rites and those of eagle trapping," each has their own individual origins and bundle lines. The Mandan and Hidatsa share many of the rites and traditions associated with Thunder ceremonies like the Big Bird, although there are some differences in traditional narratives related to the ceremonies and the transmission of bundles (Bowers 1992:362–363).

The Big Bird rites have associated bundles. Seven Hidatsa Big Bird bundles were preserved through the smallpox epidemic of 1837 and passed down through bundle lines. In 2010, a Big Bird Bundle was in the possession and use of the late Arly Knight, a Fort Berthold holy man, when we interviewed his wife, a holy woman, for this project. Big Bird bundle holders have rights to chipping flint, doctoring with flint and black medicine root, and rain-making. Black medicine root is believed to have its origins with the story of Two Men, Black Medicine and his brother Sweet Medicine, which is intertwined in the Big Bird Ceremony myth (Bowers 1992:364). In the story, Black Medicine heals a mysterious wound on an eagle child using black medicine root (a root of the red baneberry, *Actaea rubra*) and lancing it with an arrow-point (Bowers 1992:360).

One example of a complete bundle for the Big Bird rites contained

a "'Sleep feather' of the eagle; white feather from near the tail feathers of the eagle; feather from the eagle's head' feather from the left side of the eagle's claw; 12 bird sticks made of chokecherry and painted red; bullsnake skin representing the Grandfather[1] in the Missouri River; a turtle shell; an otter skin; a ferret skin; sage; a wooden sword with an image of lightning on each side; a flint knife; rattles; and a wooden pipe [Bowers 1992:364].

Wolf Chief gave an account of a bundle purchase, in which he explained it had the 12 bird sticks that represent the eagle's 12 tail feathers, and the 12 sticks used in the

"moccasin game." These could be taken into war as protection (Bowers 1992:365). The flint knife represents an eagle claw related to flint chipping and doctoring rights (Bowers 1992:368). The "sleep feather" commemorates Two Men's wives who could not return home with them except in spirit in the form of eagle feathers (Beckwith 1937:62). The sword in the bundles represents those given to Black Medicine and Sweet Medicine by the wives of the big birds in order to kill a dangerous serpent in a ravine (Bowers 2004:268; for a slightly different version, see Beckwith 1937:86; Wilson 1908:128–129).

In a variant of this story involving Packs Antelope, as told to Wilson by Good Bird, Packs Antelope is advised that he will be given a choice of wooden swords when he meets with the Thunderbird, and that he should choose the oldest and most mended, because it was the most powerful.[2] An "eagle-killing song" is also associated with the swords and the Big Bird Ceremony myth (Bowers 2004:268).

The 12 bird sticks mentioned as part of the Big Bird Bundle represent birds important to the Mandan and Hidatsa, as described by Bear-on-the-Flat to Bowers (2004:239) as:

Bald-headed eagle. White head and white tail. Parent to speckled eagle.
Speckled Eagle.
Spotted-barred tail. Body is black.
White tail with black tip. Known as the calumet eagle.
"Half of each feather good, half bad."
"Black but has white tails"; head is white; is a hawk.
I'pamasina.
Chicken hawk. Tails barred black and light brown.
Goose.
One of the smaller geese.

This list illustrates the complexities of native bird taxonomy (Bowers 2004:239). For eagles, the distinctions are based largely on plumage. The bald eagle is identified for its adult plumage, the distinctive white head and tail; it is recognized as the "parent" of the "speckled eagle," however the speckled or spotted eagle is recognized as a different kind of bird because of its plumage. A juvenile bald eagle lacks the white head and tail of an adult and is covered with irregular white markings for the first two to three years of

---

[1] "Grandfather" refers to Black Medicine after he ate part of a water serpent and changed into a snake with one head on either end of his body. Grandfather lives in the Missouri River, with one head at Bird Bill Butte (also known as Eagle Nose Butte in Morton County, North Dakota) and the other head at Thunder Creek (located in Tripp County, South Dakota) (Bowers 1992:360). He tells his brother that all offerings to the Missouri River must be made when there is no ice on the river, announces that he will be seen for the last time the next year at Thunder Creek when the Juneberries are ripe, and asks that their parents come to see him.

[2] This aligns with the idea that "the older and more complex the bundle [or its contents], the stronger the power will be to change the course of life's events; both age of the bundle and content of the originating vision are important factors in its authentication" (Zedeño 2008a:368).

Figure 9.6. Turtle Effigy State Historic Site. State Historical Society of North Dakota, photograph by Brian Austin.

life. The "calumet eagle" is generally understood to refer to the Golden Eagle, but its description of a white tail with a black tip may refer specifically to the second-year juvenile Golden Eagle whose tail has a stark black and white contrast that fades as the eagle becomes a full adult. Wilson's Hidatsa informant Wolf chief identified the speckled eagle as the "old golden eagle" (Wilson 1908:35); indeed, at full maturity the Golden Eagle's feathers are spotted under the wing and barred on the tail (Sterry and Small 2009:92).

These eagle classifications appear in oral traditions like the Hidatsa story of the Two Twins and Long Arm, where one of the twins travels through four villages in the sky world that are each populated by a different type of eagle (Wilson 1908:33). Other bird identifications also rely on feather markings.

In addition to doctoring and rain-making rights, those with Big Bird bundles also perform hunting rites at animal stone effigies that are in charge of "clearing fogs" and are located within the traditional hunting grounds of the Hidatsa (Bowers 1992:369). In one account of clearing fog for a hunting expedition, Clam Necklace (a Mandan with a Thunder or Big Bird bundle) and a group of hunters encountered a thick fog soon after they had visited a turtle effigy (Figure 9.6). They had offered flint to the turtle as a symbol of the eagle, one of seven important spirits associated with the Missouri River. Clam Necklace used a flint knife to trace an image of the turtle in the ground, and after praying to the turtle and establishing a connection to the eagle through the flint knife, the fog cleared and they were able to continue (Bowers 1992:370).

MANDAN AND HIDATSA EAGLE-TRAPPING RITES

Eagle trapping was an integral part of the Mandan and Hidatsa ceremonial complexes; in fact, many sacred bundles used to call buffalo and propitiate gardening were also used to catch eagles (Bowers 1992:138). Hidatsa eagle trapping is a practice that was likely acquired from the Mandan when the Hidatsa came to the Missouri River from the east, a landscape unsuited to eagle trapping (Bowers 1992:78, 2004:108; Wilson 1928). Their trapping traditions are so similar that Bowers describes the two tribes as if they were one when discussing eagle trapping. Although there were likely some differences in tradition, the two tribes are grouped for discussion here. Eagle-trapping rites were performed throughout the year for many purposes beyond just eagle trapping. They were also used to trap fish, corral buffalo, and for success in hunting or warfare (Bowers 2004:108).

The effects of several waves of smallpox epidemics led to a severe reduction in the number of Eagle-Trapping Bundles. The last known Eagle-Trapping Bundle is curated at the State Historical Society of North Dakota (Bowers 2004; Murray 2011). Federal regulations against acquiring eagles have also discouraged any revival of eagle-trapping practices, and today those with eagle medicine rather than trapping rights are called on to fulfill important ceremonial roles. However, eagle-trapping sites are still visible on the land and are important places to the Mandan and Hidatsa as a tangible connection to their past in a physical and spiritual landscape, as sources of knowledge and power, and as important places for gathering plants (Murray 2011; Zedeño et al. 2006:179). An examination of the ceremonial objects associated with eagle-trapping rights is worthwhile to better understand the connections of the Mandan and Hidatsa to eagle trapping as a physically and spiritually demanding process, as well as its connections with other key ceremonies.

Bird sticks are an important part of eagle-trapping rites and the Big Bird ceremony, although these two ceremonies are distinct from one another (Wilson 1928; Bowers 1992:363). Bird sticks were believed to communicate to the "scout" eagles that the eagle trappers were respecting them. The scout would report back to the other birds so that the trappers would have good luck in catching eagles (Bowers 2004:237). The Bald Eagle is known to be the scout for other eagles (Bowers 1992). The assignment of the bird sticks varied depending on the judgment of each trapping leader, but generally 12 bird sticks were stuck into the ground in a row in front of a sacred buffalo skull inside the eagle-trapping lodge (Bowers 2004:235). These were decorated with flint, a strip of rabbit skin, and white sage. In addition to the 12 bird sticks, which represented the 12 tail feathers and the birds associated with the eagle family (mentioned previously), four bird sticks were placed on the altar for each trapper to represent four types of eagles: the bald-headed, speckled, four stripes on tail feathers, and calumet eagles (Bowers 2004:235).

In addition to the bird sticks, the eagle was represented by a chokecherry twig that was attached to a hoop of twisted wheat grass, hung on a crosspiece in the back of the ceremonial eagle-trapping lodge. The wheat grass hoop was in turn representative of the Sun, once again reaffirming the cosmological connection between eagles and the Sun. A similar grass hoop hung in the front of the lodge representing the Moon, with a twig of juneberry that symbolized the "black-tipped," or calumet eagle (Bowers 2004:237). Four rocks were also brought inside the eagle-trapping sweat lodge to honor the birds. Each rock had a specific meaning: The first rock was named for "the mother of the birds," the second for "the child of the mother birds," the third for "the oldest bird," and last for the "genuine speckled eagle" (Bowers 2004:237). Inherent in the beliefs behind ceremonial objects used in eagle-trapping rites is the belief of personhood within eagles. Each step in the eagle-trapping process reflects respect for and communication with the eagle as a spirit-person who has offered itself to the trapper. Those who did not follow the eagle-trapping rites correctly ran the risk of offending the eagle and confronting the physical dangers of trapping. In Washington Matthews' account of the Hidatsa, some men's hands were "crippled for life" by the eagle talons as a result of improperly performing the rites (Matthews 1877:60).

CORN RITES OF THE MANDAN AND HIDATSA

The Corn Rites of the Mandan and Hidatsa, which are very similar to one other, are primarily linked to the agricultural practices of these groups, involving mainly corn but also squash and beans. The Old-Woman-Who-Never-Dies Bundle and the Good-Furred-Robe Bundle are two of the principal ceremonial items for agricultural rites. As mentioned previously, the Mandan trace the origins of the Old-Woman-Who-Never-Dies Bundle back to the culture hero Good-Furred-Robe, who founded the Goose Society to dance in honor of the waterbirds that come and go with the agricultural seasons, bringing rain and good luck from Old-Woman-Who-Never-Dies (Bowers 1992:338). Bowers (1992) asserts that the Hidatsa do not have any specific origin stories with the Goose Society, and thus judges that the tradition was probably borrowed from the Mandan. Regardless, the Old-Woman-Who-Never-Dies rites have deep cultural roots in both tribes and constitute an important part of traditional ceremonial and material culture.

Old-Woman-Who-Never-Dies lived on an island near the mouth of the Mississippi River in a large earth lodge (Beckwith 1937:53, 57–58; Bowers 1992:204, 2004). Migratory water birds such as ducks, geese, and cranes are associated with Old-Woman and agriculture and the changing of the seasons; they are believed to fly to and from Old-Woman's island, where they winter.

> I guess with [Grandmother Who Never Dies], all these [waterbirds] were women, the female side of the birds. We would always turn into birds and go back to her. And she would turn them into a braid of sweetgrass and take care of them in the winter in her lodge.
>
> —Mandan-Hidatsa consultant

These waterbirds carry the offerings of the Mandan and Hidatsa back to Old-Woman and in return she brings good crops for the people. Several birds have specific relationships with particular plants: The wild geese are associated with the corn, the swan with the gourd, and the duck with beans. Several other birds and animals are also associated with Old-Woman because they are traditionally known to help Old-Woman in her garden; this includes blackbirds, white-tail deer, elk, mice, and moles.

The relationship between Old-Woman-Who-Never-Dies, Good-Furred-Robe, birds, and agriculture is reflected in the Goose Society bundles. The Good-Furred-Robe Bundle of the Mandan contained a wooden pipe with a goose head carved into the stem to symbolize the growing season, which begins and ends with the arrival and departure of geese during their migration cycle (Peters 1995:112). The pipe was presumably used by Good-Furred-Robe himself in ceremonies.

> When she smokes [Grandmother's pipe] and blows that whistle it will cause water. So when she blew that pipe, and blew that whistle, she asked for the birds to help her pray. And what she asked 'em, all these women to help her pray for rain. So these birds have a big part in the rain coming. Any time you hear them screamin' or chirping, that's their cry for rain, because that's part of their world, it's where their world is.
>
> —Mandan-Hidatsa consultant

Also included in the bundle, along with animal skins, robes, and sage, are the heads of blackbirds, a duck head, and a deer skull (Bowers 2004:185). The bird and deer skulls represent the seasonal movements of the animals as linked to the agricultural cycle, and are connected to Old-Woman-Who-Never-Dies as her garden helpers; thus, the Good-Furred-Robe bundle is closely related to Old-Woman-Who-Never-Dies.

There are several groups of bundles that relate to the Mandan Old-Woman-Who-Never-Dies corn ceremonies, including principal bundles, secondary bundles, composite bundles, and sacred pot bundles. Similar to a Good-Furred-Robe bundle, an Old-Woman-Who-Never-Dies principal bundle might include blackbird heads, duck heads, swan feathers, goose heads, corn, beans, squash, sunflowers, and other skins and ceremonial objects (Bowers 2004:188). A circular drum decorated with goose tracks was sometimes included in the Old-Woman-Who-Never-Dies bundles (Bowers 1992:345). The principal bundles of Old-Woman-Who-Never-Dies rites are used for the benefit of the community and have material correlates with important people and events in the origins of the tribes. Secondary bundles, passed down hereditarily, contained many of the same items as the principal bundles, but their transfer and use were much more specific.

Mandan and Hidatsa women put offerings of meat and hide on sticks in their gardens for the birds as they flew through the Missouri River Trench in their migrations. Occasionally a woman would receive a dream from spirits of the birds or other animals associated with the garden and Old-Woman-Who-Never-Dies, and would create a new personal bundle according to their vision and hang it in the garden as a "protector" (Bowers 1992:340). These offerings once again demonstrate the role of birds as messengers between the human and spirit agents. Birds relay messages to Old-Woman-Who-Never-Dies about the personal offerings and their treatment, and Old-Woman responds accordingly by ensuring successful harvest.

CHAPTER TEN

Bird Hunting and Trapping

Traditionally, birds were an important food supplement along the Missouri River. Grouse and waterfowl, in particular, were valuable at times when other food sources were unavailable. Children hunted them opportunistically. There is a widespread story of Coyote, the trickster, who is starving and comes across a group of birds. Through mischief and deceit, he kills and roasts the birds, often only to have them eaten by another clever animal (for examples, see Grinnell 1913:185–188; Lowie 1918:33–34, 1939:209–211, 1960:11–16; Parks 1991:867–870). This story, which in its variations includes prairie-chickens, ducks, and geese, illustrates the importance of these birds to Native diets.

Not all birds were created equal in the diet of the Missouri River tribes, but as the early journals and ethnographies document, a wide range of birds were hunted and consumed. Given the spiritual significance attached to birds it is not surprising that each tribe had unique protocols for hunting and eating birds. A Blackfoot consultant, for example, observed that duck eggs were sacred food that were consumed during bundle openings and other ceremonies. Today, this tradition continues with commercial eggs. Native proscriptions almost universally prohibited consumption of sacred birds, particularly eagles, because of their status as important spirit persons and as Thunderbird on earth.

Through the analysis of archaeological assemblages, Falk (2002, also Falk et al. 1991), Parmalee (1977b, 1979), and Ubelaker and Wedel (1975) have contributed greatly to the understanding of protohistoric and historic uses of birds in subsistence, social, and religious contexts. Parmalee (1979) examined the animal remains from the Mobridge site (Johnson 2007:145, 191–194), a Post-contact

(A.D. 1600—1700) Arikara village located along the Missouri River in South Dakota. Study of the faunal materials yielded the presence of 17 different species of birds, accounting for almost half of the number of animal species present at Mobridge, although bird bones represented just less than 5 percent of the total animal remains (Parmalee 1979:197). Not all of these species were edible. Although waterfowl, grouse, raptors (eagles, hawks, owls, etc.), and corvids (ravens, crows, magpies, etc.) were all hunted, grouse and waterfowl supplied most of the dietary functions while raptors and corvids had "primary value" in their social and ceremonial uses (Parmalee 1979:197). Parmalee further notes that the avian assemblage from the Mobridge site is consistent with "a pattern of bird utilization characteristic of most prehistoric middle Missouri Valley groups," suggesting that the significance and use of these birds spans back well into the prehistoric past.

Likewise, at the Scattered Village site in Mandan, North Dakota, Falk (2002) encountered a higher than expected frequency of bird remains in contexts dating to just before A.D. 1600. That higher frequency continued for perhaps a century, but no later than A.D. 1700 (Ahler 2002:18.3). It is unclear whether this site was occupied by the Mandan, Hidatsa, or another group, but the Mandan seem the likely owners of this site, notably because Red-headed Woodpeckers were found and the Hidatsa never killed this species (Ahler 2002; Mitchell 2013; Weitzner 1979). From 1,786 bird specimens Falk was able to identify 34 species representing 10 orders and at least 15 families. Of particular interest were the 20 woodpecker specimens including the remains of 7 Red-headed Woodpeckers, 4 Northern Flickers, 1 Downy woodpecker (*Picoides pubescens*), 1 Hairy Woodpecker (*P. villosus*), 1 Yellow-bellied

Sapsucker (*Sphyrapicus varius*), and 6 woodpeckers of indeterminate species. All of the identified species occur in the upper Missouri Basin (Falk 2002:7.19). Falk notes that high bird bone frequency and species diversity are common in Middle Missouri archaeological sites. Parmalee (1980) also found a relatively large sample of Red-headed Woodpecker remains at the Bagnell site in Oliver County, North Dakota. Although few in number, these assemblages present a diverse and compelling picture of people's ability to obtain birds for various uses.

TRADITIONS OF BIRD HUNTING

Tied into the dietary value of birds are the traditional processes associated with hunting as well as the cultural values that are taught and reinforced through the act of hunting for subsistence. Birds were used by many tribes to teach the elementary skills of hunting to young boys, which would be applied later to hunting larger game that contributed a larger portion of protein to the diet. An "important activity for young boys," bird hunting also was accompanied by certain stories, songs, and customs that reinforced tribal beliefs and identity (Gilman and Schneider 1987:72).

According to Blackfoot tradition, Old Man Napi, the creator of the earth, plants, animals, and humans, taught the Blackfoot how to protect themselves from bison who had been killing and eating the people. As "master of all the birds," Napi easily caught a bird and plucked its feathers, making the first arrow out of serviceberry, black flint,[1] and split feathers tied to the shaft of the arrow (Grinnell 1920:140). He taught the people how to use these things to hunt bison and prepare it to eat. Such a story might be shared between hunters and the young boys as they made their first attempts at hunting birds.

Shooting small birds was one of the ways that young boys began to learn how to hunt and use a bow and arrow. When a boy killed his first bird in the presence of a clan aunt, according to Hidatsa traditions, she had the opportunity to take the meat for herself. Witnessing the first successful hunt was considered a service, and she was compensated by the boy's father with a horse or some other kind of gift (Weitzner 1979:199). Thereafter, a bird might be given to a boy's mother or grandmother, who would roast the bird whole until the skin and feathers fell off and

eat it as a delicacy (Weitzner 1979:197; Wilson 1911:104). In other cases, bird meat was given to elders, because it was easier to chew than other meats (Gilman and Schneider 1987:72). In an interview with Claude Schaeffer, Piikani elder Ms. Yellow Kidney recalled that during her childhood young boys would shoot small birds with a bow and arrow and bring them back to camp for the young girls to prepare (Schaeffer and Schaeffer 1934: GM, Blackfeet M-1100-143). She remembered boiling the small birds in tin food cans hung from a tripod over an open fire.

Certain etiquette also accompanied bird hunting among the Hidatsa. Birds were snared or shot with a blunt-headed arrow in order to kill the bird without destroying its flesh (Gilman and Schneider 1987:72; Weitzner 1979:198 and 239–240). When hunting with bow and arrow, the hunter who picked up the bird first was the owner, whether or not he was the one who killed the bird (Weitzner 1979:197). However, if a bird was snared, it belonged to the snare's owner. Hidatsa elders would set up snares to catch Snow Buntings in the winter with the help of young boys; as they set the snares, the elders would sing "My snare come to, hey, hey!" or "Hey, hey, come to my snare!" (Weitzner 1979:198). The boys would observe the elders as they set up the snares and sang the song, and later would chant "Snare-twisted face" as they watched the snares, waiting for the buntings to approach. Then, the boys would rush out to the snares to claim the first bird they reached.

Bird hunting was a pastime. As they tended gardens, Hidatsa boys were known to hunt blackbirds and "moccasin birds" (Weitzner 1979:197). Wolf Chief listed more than 15 different types of birds that were hunted by young boys, categorized by the environment they could be found in (the woods, hills, bottoms, gardens) and the season (Wilson 1911:98–99). Most of these birds were identified in Hidatsa, and their species cannot be surmised, but the variety alone is notable. Crow boys might split into two groups and compete to see which team could shoot the most birds in a span of time. The losers were "made to chew the feet, the legs, and the wings," (Lowie 1922a:250). Wilson (1924) also remarked that young Hidatsa boys would hunt and eat blackbirds "to pass the time" while they minded the horse herds.

Edible Birds

Gallinaceous birds are generally a nonmigratory group, and therefore were available year 'round as a supplementary food source. Prairie-chickens and grouse were among the edible foods and thus hunted by everyone along the Missouri River (Figure 10.1; Bowers 2004:30;

---

[1] Grinnell's (1920:140) account mentions black flint as making "the best arrow points, and some white flints," suggesting white flint is also useful as a secondary source for arrowheads.

Figure 10.1. Wendi Field Murray and Mandan elder at a grouse mating ground and hunting place in North Dakota (2012). Photograph by M. N. Zedeño.

Hungry-Wolf 2006:143; MWP 1942:164–165; Parmalee 1979; Weitzner 1979:197; Wilson 1928:242). According to Wolf Chief, prairie-chickens were not an important food source for the Hidatsa in the "old times," and were therefore more abundant in the past (Wilson 1928:240). Prairie-chickens are known for their mating dance, and they were considered "too wild" for young boys to hunt, along with ducks and geese (Weitzner 1979:197; Wilson 1911:106). They were typically snared, and sometimes the chicks were taken and raised in captivity by the Hidatsa until they eventually escaped (Wilson 1928:242).

Although their identity is tied to the prairie-chicken, the Prairie-chicken Clan of the Mandan did not have any spiritual prohibition that kept them from hunting or eating this bird (Bowers 2004:30). Personal taboos about killing the bird may have arisen from individual experiences and dreams. Grouse and prairie-chicken feathers or skins were occasionally used in bundles and other objects of spiritual significance. However, their use in such bundles is sometimes associated with extraordinary circumstances, such as a Peigan Beaver Bundle that contains a "white prairie-chicken" (Wissler 1912:170), which is likely an instance of a rare albino form (Parmalee 1977b:211).

Such cases take on special significance and meaning for their anomaly.

Prairie-chickens have "declined catastrophically" since the westward expansion of settlement, overhunting, and destruction of prairie habitat (Sterry and Small 2009:58). The Sharp-tailed Grouse (*Tympanuchus phasianellus*) is still locally common, although it has similarly declined due to reduction and degradation of habitat (Sterry and Small 2009:60). Competition for habitat with the reproductively successful, non-native Ring-necked Pheasant (*Phasianus colchicus*) has also contributed to their near demise.

Migratory waterfowl also constitute a large part of the avian diet of Plains tribes; although they are only available during part of the year. Main "staging areas" —places where birds stop to rest and feed along the flyway—provide opportunities for hunting waterbirds. A study of 51 archaeological samples from sites in South Dakota documented at least 13 different species of waterfowl in the archaeological record (Parmalee 1979a:200), possibly more considering that some bones could not be differentiated at a species level. Swans, geese, and ducks constituted 15 percent of the total number of avian remains at these sites (Parmalee 1979a:200).

The seasonal hunting of waterbirds began in March, when people would gather "along the lakes and rivers to build their canoes, trap furs, fish, and hunt waterfowl" (Ray 1974:46). Edible waterbirds include cranes, ducks, geese, and pelicans (Denig 2000:189). Eggs were collected and cooked in early Spring. Grinnell documented the "great feasts of the eggs of ducks and other water-fowl" that the Blackfoot celebrated in the spring (Grinnell 1920:207). According to a Blackfoot consultant, in their traditions duck eggs are "mystery food" or "magic food" with special significance for Beaver Bundle holders.

In Summer, the Blackfoot would hunt young ducks and geese who were not yet able to fly by ambushing them at the shallow lakes where they gathered (Grinnell 1913:208–209); fire was used to drive ducks into the ambush. Likewise, Maximilian's journals corroborate the wealth of waterfowl in the summer months, and the vulnerability of the birds to hunting during this season, remarking that, "the old geese lie down and pretend to be disabled because of maternal love. They are molting now, and some cannot fly…" (Witte and Gallagher 2008:216). Schaeffer also reports that older ducks were taken near lakes by the Blackfoot during the molting season (Schaeffer and Schaeffer 1934: GM, Blackfeet M-1100-143). The Assiniboine also hunted ducks during the Summer (MWP 1942:165).

There is an interesting relationship with waterfowl and subsistence practices for the Mandan, Hidatsa, and Arikara. Waterfowl are at once an important source of food in the Spring, Summer, and Fall, but during these same seasons they are also intricately linked to planting and harvesting. Offerings and prayers are made to the waterfowl to ensure their return because of their significant role in the agricultural cycle and perhaps because they provide such an important source of protein. Waterbirds were hunted not only for food, but also for their feathers and other parts that were incorporated into objects and regalia used in planting and harvest rites.

In addition to waterfowl and grouse, a number of other avian species contributed to Native diets in the Missouri River Basin. Information from ethnographic sources provides early historical accounts directly from Native hunters and consumers, while the uses of those found in archaeological contexts can be speculated on. The species that occur in archaeological contexts often closely align with historic records from Native people, suggesting that subsistence uses have largely remained the same over time in this region.

For example, Sandhill Crane bones made up 6.8 percent of the bird remains from the Mobridge site, where geese and ducks made up 13.5 percent and Sharp-tailed Grouse constituted 20.5 percent of avian bones (Parmalee 1979:198). Several Assiniboine sources mention the crane as a bird that was considered good to eat (Denig 2000:189; Rodnick 1938:25). The number of crane elements found in the 51 Arikara sites included in Parmalee's study of Plains avifauna "suggest that these birds were a valued food resource to many groups" (Parmalee 1977a:212).

Consumption of other bird species varied by tribe. For example, Denig reports that "the Assiniboine were known to eat all birds, even the crow" (2000:93), while Mandan sources are almost completely devoid of any reference to the Mandan eating birds. It is difficult to compile an exhaustive list of the birds that were hunted and eaten by each tribe, as little attention has been given to birds in native subsistence compared to large game such as bison, deer, and pronghorn. Assiniboine and Hidatsa records contain the most information about how birds fit into subsistence practices. Drawing from the available commentary on birds as food in ethnography we can begin to gain perspective on which birds (after waterfowl and grouse) were most often acceptable for consumption.

Although Denig purported that the Assiniboine might eat "all birds," those specifically noted in the text include cranes, crows, ravens, magpies, owls, and pelicans, while eagles are excluded. Eggs were also commonly eaten, except for eagle, crow, and magpie eggs because of their unappetizing flavor (Rodnick 1938:27). The Hidatsa hunted and ate a wide range of birds, including blackbirds, bobolinks, cowbirds, a crossbill species, Mourning Doves, Northern Flickers, Snow Buntings, and various woodpeckers (Weitzner 1979; Wilson 1924). As a general rule, birds of prey were not eaten, although the Hidatsa would sometimes eat them when other food sources were scarce (Matthews 1877:24). Many of the birds recorded by early ethnographers as typically hunted remain so generally named or described that it is hard to assign a species to them: "Small brown bird," "bird-yellow," "breast-yellow," "small cherry," and "grape-eater" are just a few mentioned in Wilson's notes on the Hidatsa (Weitzner 1979:197–198). Therefore, the list of edible birds remains incomplete.

Non-edible Birds and Personal Taboos

Some birds were not eaten simply because of aversion to the taste of their meat. Schaeffer specifies discrimination among the Blackfoot for eating grebe or loon, apparently because of their "coarse and unpalatable" taste (Schaeffer and Schaeffer 1934: GM, Blackfeet M-1100-143). For some people, however, taste was not the issue. Their spiritual significance overrode their value as a potential food source. Such is the case of eagles. According the Blackfoot, eagles were respected as "sacred, very wise, and almost human-like," and therefore their flesh was never eaten (Hungry-Wolf 2006:137). This was common for other Plains tribes as well. Eagle trapping was a dangerous and sacred process related to the Thunderbird. The Hidatsa did not eat white-breasted swallows because their forked tails resembled that of the Thunderbird, and therefore they were thought of as Thunderbirds and too sacred to eat (Weitzner 1979:312).

Magpies and wrens were too sacred to the Hidatsa to be consumed, although magpies could be shot for their feathers (Weitzner 1979:312). Similarly, hawks were shot for their feathers but were not eaten. Meadowlarks were not considered a food species by the Hidatsa, but they are believed to cure deafness and dumbness when eaten (Weitzner 1979:197). Red-breasted swallows were not consumed either, because they were thought to cause an inability to urinate (Wilson 1911:101). Weitzner specifies three types of woodpeckers that were never killed by the Hidatsa: "the one with the red head" (the Red-headed Woodpecker), "the yellow,"[1] and "the winter woodpecker with black spots" (unidentified).

In addition to the broad guidelines that regulated what was and what was not edible, personal experiences also contributed to what birds an individual might consider edible. These personal taboos were conferred on an individual in a dream or vision of their spirit helper who instructed them to abstain from eating a certain food. For example, someone who has the Crane as their medicine might abstain from eating crane meat because of their spiritual ties to that bird. In other cases, a bird may warn someone against eating a certain food and a personal or familial taboo may be created. Such was the case for the grandmother of Crow medicine woman Pretty-Shield, who was warned by a chickadee that neither she nor her family should ever eat eggs lest they be killed by the enemy (Linderman 1972:159–160). Personal taboos against eating birds and bird products were strictly adhered to in order to avoid sickness, bad luck, or in some cases, death. Taboos were lifted in cases of emergency, such as extreme scarcity of food on a hunting trip.

---

[1] Assumed not to be the yellow-shafted Northern Flicker, which is probably the "golden-winged woodpecker" referred to in Wilson's notes as a bird hunted by the Hidatsa.

EAGLE TRAPPING

Eagle trapping played a large role in the spiritual undertakings of many tribes around the Missouri River, including the Mandan, Hidatsa, Arikara, Crow, and Blackfoot as well as in their voluminous trading economy. Eagle trapping was not practiced by the Assiniboine, although eagle feathers and other parts were still highly valued because of the significance of the eagle in the Assiniboine cultural belief system. Native American acquisition of eagles and eagle feathers has been restricted by environmental legislation that aimed to protect eagles as an endangered species, such as the Bald and Golden Eagle Protection Act of 1940 and the Endangered Species Act of 1973. As Murray (2011) points out, however, the delisting of the Bald Eagle as an endangered species "has breathed new life into how extant eagle protection legislation will be implemented, and has important implications for both the management of eagle trapping sites and indigenous efforts to revitalize eagle trapping practices" under the protection of the American Indian Religious Freedom Act ([AIRFA] 42 USC § 1996) and Religious Freedom Restoration Act ([RFRA] 42 USC § 2000BB). Eagles continue to be significant as "a singularly unique resource" to the tribes, for their transcendent and transformative qualities linking the human and spirit worlds, and their ability to influence human behavior (Murray 2011).

Eagles and eagle trapping are fundamentally tied to the cultural landscape of the Missouri River and surrounding environment. Trapping had a dual purpose: to secure feathers and eagle parts needed for spiritual undertakings, and to fulfill a community's need for supplies of bison meat and hides (Wilson 1928; Bowers 2004:250). Eagle trapping sometimes also occurred in areas of rougher terrain that were not suitable for other hunting. Eagle-trapping territories were initially divided by moieties. Later, trapping rites and territories were passed down through tribal lineages such as clan lines among the Hidatsa, or from father to son among the Mandan (Bowers 1992:79). The division of the Mandan and Hidatsa tribes into moieties is reflected in the mythology of eagle-trapping camps. Eagle-trapping instructions were received from the black bears and the brown bears, living on the east and west sides of the Missouri River respectively, who came together in the fall at mythological eagle-trapping camps (Bowers 2004:215).

Wolf Chief shared a story about the gift of eagle trapping to the Mandan people as intertwined with how they learned to trap fish as well (Wilson 1909:46). Both of these skills were acquired from the black bears in this story, after a young man was adopted by the bears and had returned to his own village. The bears came to the young man's village to visit him, and the young man put on a ceremony in honor of them. It was during this ceremony that the bears taught the young man about eagle and bird trapping, as well as how to trap fish.

> When the four nights ended his fathers, the black bears, approached the village. . . . The bears advanced toward the village, stopping on the way four times. Each time they stopped they sang the mystery song. . . . After the fourth song they came into their son's house—the chief's son's earth lodge. There they sang many songs, feasted and sang again. They taught their son and feasted and ate and sang all night. They had arrived at the village just after sunset.
>
> In the morning they divided among them the skins their son had given them. As each bear received his gift of skins he taught his son something of trap or snare craft. One would say, "I will show you how to make the eagle hunters' pit." Another would say, "I will show you how to catch eagles when they are flying in the air. It is not hard to do it. I can do it and I will show you!" Others taught him how to make bird traps: how to make spring snares that can catch birds—even eagles! [Wilson 1909:46–47].

For the Crow, the buttes around Theodore Roosevelt National Park are significant in part because of their associations with the origins of eagle trapping; "Elk, eagle, and black bear, who taught people to trap eagles, live there" (Zedeño et al. 2006:241). Eagle trappers strictly followed family claims over trapping territory. Although territories have shifted over time due to migrations, political pressures, and changes in the landscape, these traditional codes of behavior continue in modified forms today. One Fort Berthold consultant explained that eagle-trapping pits represent the sacrifice that the eagle makes in order to give the tribes an important ceremony (Murray 2011). Bowers (2004:207) comments that "thousands of depressions marking the sites of former pits" can be seen west of the Missouri. Although eagle-trapping pits are no longer used for trapping purposes because of the restrictions on eagle taking, they are still sites of cultural importance to the tribes for other reasons, including vision quests.

Eagle-trapping sites continue to hold significance for Native people today because of the relationship they represent with eagles and eagle spirits, the genealogical ties to trapping territories, and their value as sources of knowledge and power. In oral traditions, eagle pits are common places for a man to receive visions and dreams with knowledge from the eagle or another type of bird:

> There is another painted lodge known as the Crow Lodge. It came about in this way. One man was catching eagles on a hill. He had made a hole in which he was hiding. After a while he went to sleep. He dreamed that a Crow came to him saying "This is my lodge. Now I shall give it to you with the medicine and songs." So the Crow transferred the lodge to the man, taught him the songs and the ceremony [Wissler and Duvall 1908].

In Grinnell's time among the Blackfoot, he observed that it was "considered improper" to hunt eagles with a bow and arrow; in any case, one could rarely approach an eagle close enough to kill them in this way (Grinnell 1920:236). The standard pit method of eagle trapping was closely bound to a spiritual undertaking between the hunter and the eagles. Pits were often located along wooded streams on the bluffs or bench lands located above the waterways. The landscape was integral to the success of eagle trapping, attributing additional significance to the Missouri River, its tributaries, and the surrounding topography. The trapper built a pit and covered himself with branches for camouflage. Bait (often meat from a rabbit or coyote) was placed on top of the branches. When the eagle alighted on top of the branches, the eagle catcher would grab the eagle by its feet and drag it into the pit (Zedeño et al. 2006:244). The eagle was killed immediately and the feathers removed in the pit; or the bird was brought back to the camp alive and the feathers removed there. This depended on the tribe and the trapping leader (Bowers 2004:246; Matthews 1877:60; Thwaites 1906:382).

Eagle trapping was a dangerous endeavor undertaken only by those who had received power or instructions from an eagle spirit and were brave enough to participate. For the Arikara, eagle trapping was second only to bravery in battle as a means of gaining honor (Parks 1996:282). The wounds obtained through eagle trapping could easily lead to fatal infections, and one account attested that Bald Eagles are so strong that "they have almost lifted me (an eagle trapper) out of the pit" (McClintock 1999:429). It was not uncommon for eagle trappers to have their hands "crippled for life" by scratches received while eagle trapping. This was considered by the Hidatsa to be a direct consequence of mistakes made by the trapper while performing the eagle-trapping rites (Matthews 1877:60).

In addition to correctly performing the eagle-trapping rites, certain traditions and behaviors were enacted by eagle trappers in order to ensure their safety throughout the expedition. For example, in the Blackfoot tradition, an eagle trapper's wife could not use an awl while her husband was away because it would cause the eagle to wound the catcher (Grinnell 1920:238). Similarly, a man with eagle-trapping power would not use a knife while eating for the same reason (Hungry-Wolf 2006:137). The eagle trapper would bring a human skull, a long stick, and pemmican into his eagle-trapping pit for protection (Grinnell 1920:237–238). It was believed that the spirit of the person's skull would protect the trapper against harm and that it would also make the trapper invisible to the eagle. The stick was used to drive away smaller birds, as well as eagles that were too dangerous to catch. Pemmican was placed in the mouth of an eagle after it was killed by the trapper. This made the other eagles hungry. Similarly, in a Mandan and Hidatsa song for eagle trapping recorded by Wilson (1909:125), one stanza says "I give you sunflower seed and corn in a ball," referring to a similar custom of offering the dead eagle food after it is trapped.

The trapping process also required a considerable amount of spiritual and physical tribulations beyond the dangerous aspects of the act. Eagle trapping required men to "suffer for the birds," and formal fasting and torturing were a central feature of trapping, especially for young Mandan men (Bowers 2004:243). These actions ensured the success of a trapping expedition because personal sacrifice would please the eagles. In fact, in the Arikara origin story of eagle trapping, the eagle expressed his and the other birds' happiness at his sacrifice and good fasting and rewarded him with the knowledge of eagle trapping (Gilmore 1929).

At this point, it is appropriate to address what can appear to be a contradiction: that humans hunt and kill birds like eagles, while also believing in the personhood, power, and agency of these birds. A common explanation of hunting relationships between humans and animals in Native worldviews lies in the idea of "hunting-as-reciprocity" (Nadasdy 2007:27). Animals are agents who "give" themselves to hunters and are spiritually compensated through the performance of rituals, offerings, and prayers to that animal by the hunter (Nadasdy 2007:25). This reciprocity may seem to be contradicted by some aspects of hunting, like the implication that at times animals must be tricked or outsmarted. This is a common motif in oral traditions. There is risk of harm or punishment when a hunter does not adequately fulfill their ritual obligations to the animal, as mentioned in regard Hidatsa belief that wounds from the eagle were the result of incorrect performance of eagle-trapping rites. This creates what Nadasdy (2007:28) terms "two contradictory principles: a positive principle

of reciprocity and a negative principle of domination"—a false dichotomy.

Reciprocity in hunting, the process of mutual exchange between humans and animals, does not necessitate altruism. In Native worldviews, human relationships with animals go beyond the caloric values or symbolic importance of animals to encompass a sociality. "When we view the human-animal relation in terms of reciprocal exchange between two agents, it becomes a "social act [that] binds persons to one another through the creation and maintenance of social relations" (Nadasdy 2007:29). Therefore, the "tension inherent in the gift relationship" is not a schism in the relationship between humans and birds, or humans and other animals, but can be viewed as a "coherent whole" of social relations between these groups.

Eagle trapping is prohibited by the Bald and Golden Eagle Protection Act, even though the Bald Eagle is no longer listed on the Endangered Species list. Permits are available from a regional Migratory Bird Permit Office to request eagle feathers and other parts from the National Eagle Repository in Denver, Colorado. This has disconnected native people from the cultural *process* of eagle trapping, which has in some ways lessened the power of eagle feathers. One consultant explained that it is "important to know the *intention* behind the feather" (emphasis added, Murray 2011); in other words, the process and intent behind the trappers' actions significantly impact the power possessed in an eagle feather or part.

Because trapping is no longer an option, the primary emphasis on rights to eagle materials has shifted from those people with the rights to trap eagles, to those who have the rights to eagle medicine. People with the rights to eagle medicine are the only ones who can apply for a permit to request eagle materials from the National Eagle Repository. Along these lines, eagle medicine rights are still sustained in cultural traditions (Murray 2011; Blackfoot consultant, 2011). The transfer and maintenance of eagle medicine, its bundles, and feather handling rights continue in ways similar to those observed by early ethnographers, although trapping rights are no longer viable. In fact, one Mandan-Hidatsa consultant noted that nowadays there are not men strong enough and courageous enough to attempt this feat. The value of eagle medicine is reinforced through oral tradition and enactment of traditional practices, and those who have eagle rights hold a special place in the community.

Eagle-trapping pits cannot be viewed in isolation from their connection to other elements of the cultural landscape, including trapping lodges, sacrificial sites, camp trails, and storied places mentioned in the origins and oral traditions surrounding eagle trapping (Murray 2011). All of these types of sites are important for their associations with eagle trapping traditions. Most stories of eagle trapping or visions occur on a hilltop setting. Cliffs, hills, or ridges near water are the preferred habitat for eagles and these elevations also place them closer to the supernatural beings above. This type of landscape is vital to eagle trapping and interactions with eagles. In addition, a Mandan-Hidatsa consultant explained that plants gathered in the vicinity of such sites are more powerful because of their location in relation to these significant cultural sites and because of the connection with tribal ancestors (Zedeño et. al. 2006:196).

The social relations between eagles and humans may further encompass the bundle of other meanings and associations that make the eagle an important part of native culture for the tribes around the Missouri River. This "bundle" of collective meanings forms what Murray (2011) calls the "eagle complex," which ties together the many associations that eagles take on in contemporary practice and understanding. Working from interviews with the Mandan, Hidatsa, and Arikara, Murray (2011) developed an "eagle complex model" expressing the contemporary connections between eagles and native religion. These connections include eagle sites (pits, camps, trails, etc.), ritual and practice, spirituality and well-being, individual and cultural identity, associations with the past and ancestors, active ceremonies and bundles, knowledge, and associated flora and fauna. Each component is interconnected with eagles, humans, and various other bird and non-bird elements, illustrative of how intricately interwoven the significance of birds is with natural resources, landscapes, and cultural resources. One day, perhaps, the elders hope there will be a generation of young men with the courage and faith needed to trap eagles in the old way.

CONTEMPORARY HUNTING EXPERIENCES

Bird hunting is still an important part of subsistence practices today, and it also provides a connection between the present and the past through the traditions and experiences shared across generations related to hunting. Hunting is a source of "enjoyment" for Native groups around the Missouri River, but also fulfills a "necessary" role in the diets of many Native families who depend on hunting to fulfill their most basic needs (Fitzgerald 1991:45). Yellowtail recounts some of his more humorous and extraordinary hunting stories to Fitzgerald (1991), such as how he "shot" a ring-necked pheasant with a fishing pole on one occasion and killed 33 ducks with just 3 shots from his shotgun on another. Tied to these experiences are the values and attitudes about hunting and the relationships that people should have with game and with the environment.

All of these stories are fun to hear, and they bring back memories of wonderful times. As we consider the fun of hunting, we should also remember the importance of hunting to the Indians and the attitudes with which each hunter should approach his goal. Indians of olden days depended on hunting for their survival, and it was almost an everyday occupation. . . . Today, we still need meat, because it is difficult to afford the basic things we need to survive [Fitzgerald 1991:45].

The importance of hunting to Native American tribes is reasserted in the treaty rights that were ensured to these groups when the U.S. government moved them onto reservations during the nineteenth century. In signing treaties like the 1851 Fort Laramie treaty, the Crow, Assiniboine, Mandan, Hidatsa, and Arikara retained the "privilege of hunting, fishing, or passing over any of the tracts of country heretofore described" that were traditionally used by those tribes (1851 Treaty of Fort Laramie, Article 5). For the Blackfoot, these rights were protected in the 1855 Lame Bull Treaty (Ewers 1974).

The continuance of traditional resource use in places that the tribes used for millennia fosters an important link to tribal sovereignty. Hunting birds and other game contributes to the basic cultural, physical, and spiritual needs of the community because it serves to replenish bundles and other ceremonial objects as well as to feed those who cannot afford their grocery bills. But most importantly, it serves to preserve community identity and individual freedom.

CHAPTER ELEVEN

A Future for the Society of People and Birds

When Robert Lowie recorded the Crow Sun Dance in 1915, it had not been practiced for more than 40 years because of the prohibition on many traditional Native American practices and ceremonies during the transition to reservation life. In 1882 the Religious Crimes Code was established by the Commissioner of Indian Affairs to stamp out Native American religious practice under penalty of imprisonment and withholding of rations (Prucha 1995). Lowie (1915b:17) explained that the Crow Sun Dance was performed only when a person pledged the ceremony in mourning for the death of a relative in battle. This close link to warfare also hastened its decline. With the move to reservations, warfare between tribes became a thing of the past.

The federal policy of suppressing traditional Indian ceremonies and dances was officially withdrawn by John Collier in 1934 with a directive called *Circular No. 2970: Indian Religious Freedom and Indian Culture*. But the Crow Sun Dance traditions had been lost over the years. The modern Crow Sun Dance has been adapted from the Shoshone Sun Dance tradition, which secretly continued to be practiced through the period of its outlaw and was taught to the Crow by John Trehero, the Shoshone Sun Dance chief, in the early 1940s (Medicine Crow 2007). Thus, some of the practices Lowie recorded from an earlier version of the Crow Sun Dance may be altered in their present forms to more closely represent the Shoshone traditions. For example, the early, reservation-era Crow Sun Dance was not originally an annual ceremony, but was only practiced when someone pledged to have a ceremony in order to seek revenge for a slain loved-one. In its present form, the Sun Dance is held yearly in the Spring or early Summer after the first thunder (Fitzgerald 1991:136).

The recorded history of this ceremony and of numerous others demonstrates that North American ethnography, as it was conducted in the late nineteenth and early twentieth centuries, has a critical, although admittedly controversial, place in the survival of Native American cultural and religious practices. Descendants of those who suffered through cultural eradication efforts have learned the bitter consequences of not passing down cultural knowledge to future generations. Interviews with leaders and elders of both sexes; maps; portraits; landscape paintings; photographs of objects, places, and practices; musical recordings; and detailed written reports all have currency in tribal efforts to revitalize traditional culture. Jerry Potts, a North Peigan religious leader, for example, describes John Edward Curtis's photographs of the Blackfoot Okan (Curtis 1997) as "inspiring for our people" because they aided in recreating the ceremony in the twentieth century. Blackfoot consultants tell how the discovery of Frances Densmore's musical recordings of the Tobacco Planting ceremony allowed Blackfoot elders to relearn the sacred songs. Society leaders and bundle holders are bound to secrecy by their office and thus cannot reveal certain information about birds and other resources to an outsider. Yet, many of them have scrutinized the ethnographic literature and thus regularly point out to the authors those sources that provide reasonably accurate versions of their stories and traditional practices.

Likewise, Calvin Grinnell, a Hidatsa Sun Dance elder and tribal historian, regards the ethnographic work of Gilbert Wilson and Alfred Bowers as testament to the foresight of his ancestors, who shared their vast knowledge with these ethnographers so that future generations would not lose their religion. Grinnell attributes the survival of

Figure 11.1. Blackfoot consultant at an eagle nesting site (2013). Photograph by M. N. Zedeño.

his people to the preservation of this religious knowledge (Murray et al. 2011). Crow elder and Sacred Tobacco Society leader George Reed also points out that ethnography and archaeology, although sometimes intrusive and inaccurate, do bring back a connection between people and the past. This connection is sometimes very personal, as Assiniboine elders found when they saw portraits of their ancestors dressed in ceremonial regalia in Karl Bodmer's studio art (Hunt et al. 2002). These heartening views of an oft-maligned profession serve to temper contemporary criticism by Indian and non-Indian intellectuals (e.g., Biolsi and Zimmerman 1997; Meyer and Royer 2001).

Twenty-first century ethnography has taken a decidedly different and positive turn, as North American anthropologists are deeply committed to ethical research that requires engaging tribal communities as collaborators who can influence and even veto research topics and questions being asked of them (e.g., Colwell-Chanthaphonh and Ferguson 2008; Posey 2004; Stoffle et al. 2001; Zedeño 2008b, 2014). It is in this spirit that we undertook the research for this book, and that tribal consultants agreed to let their voices be heard in this study, so that the

society of people and birds may continue its trajectory and strengthen its bonds, despite culture loss and the clear and present threat to bird habitats.

Knowledge of birds and the practice of using bird spirits and bird parts in daily cultural and ritual practice has changed since ethnographers first entered Native communities, often dramatically so. Yet birds are present and thus continue to interact with people. The repatriation of sacred bundles has had a major effect on the revitalization of practices that, among other things, reconnect people and the power of birds (Conaty 2015). Perhaps the range of bird species and the practices associated with certain birds has shrunk from its pre-contact breadth, but the birds that European explorers and early anthropologists recorded as being the most influential in cultural practices continue to occupy a prominent place in Native American life.

Raptors, songbirds, waterbirds, and others figure in contemporary stories just as they did in the past. They convey messages from above, warn people of danger, and bring good luck. Furthermore, birds are a means to connect tribal and mainstream culture in America. Consider the popularity of turkey at every household table—a celebratory bird that proverbially helped establish the first connection between two vastly different cultures. The turkey was, for some woodland tribes, equivalent to the eagle in cultural significance. The turkey has unique symbolic connotations, both good and bad, for all Americans.

Just as the dove may have saved the life of Francis Chardon in 1837, birds have helped the authors in many ways. The sight of an eagle flying above during an interview, the company of a magpie, or the song of an unseen bird influenced the character of the conversation, the information conveyed to the ethnographer, and even the depth and quality of a professional relationship. The presence of certain birds has an important impact on Native American assessments of cultural resources. For example, nighthawks nesting at an archaeological site remind the Blackfoot of the connection between these birds and "Old Man Napi and the Rock," thus increasing the site's cultural significance. An eagle nest can literally bring life to an old site, while an eagle-trapping pit consecrates a place and a cluster of them makes a landscape holy (Figure 11.1). When archaeologists and ethnographers offer tribal members the opportunity to visit sites and examine important resources that may no longer be readily accessible in reservation lands, they can further strengthen professional

collaborations *and* deeply enrich reconstructions of the past. Furthermore, these exchanges help to bring together resource managers and tribal cultural practitioners who require birds and bird parts for replenishing their bundles and revitalizing traditional religious life.

Is there a future for the society of people and birds? The society is alive in its contemporary form, as stories of modern warriors and birds, among others, demonstrate. Its ancient, more traditional expressions persist in some cases, as in the Blackfoot Beaver Bundle and Thunder Pipe, or the Crow Sacred Tobacco Society, whereas in other cases—the Mandan Okipa—old expressions have disappeared in practice, but have not diminished in memory or significance. As long as birds find a suitable environment for themselves and their offspring and people are free to practice their culture and religion, people will maintain and feed this intricate relationship with stories and practices that not only recapitulate tradition but also imbue the society with new narratives and experiences. Scientists, ethnographers and Native Americans can continue to find common ground where conservation and culturally informed practice meet, and where multiple voices can blend in a victory song.

References Cited

Abel, Annie Heloise
    1997   *Chardon's Journal at Ft. Clark: 1834–1839.* Lincoln: University of Nebraska Press.

Ahler, Stanley A.
    1986   *The Knife River Flint Quarries: Excavations at Site 32DU508.* State Historical Society of North Dakota: Bismarck.
    2002   Summary and Conclusions. In *Prehistory on First Street NE: The Archaeology of Scattered Village, Mandan, North Dakota,* edited by Stanley A. Ahler, pp. 18.1–18.5. Submitted to the City of Mandan and North Dakota Department of Transportation, Bismarck. Flagstaff, Arizona: Paleo Cultural Research Group.

Ahler, Stanley A., Thomas D. Thiessen, and Michael K. Trimble
    1991   *People of the Willows: The Prehistory and Early History of the Hidatsa Indians.* Grand Forks: University of North Dakota Press.

Audubon, Maria R.
    1960   *Audubon and His Journals,* Vols. 1 and 2. New York: Charles Scribner's Sons.

Barbour, Barton
    2001   *Fort Union and the Upper Missouri Fur Trade.* Norman: University of Oklahoma Press.

Bastien, Betty
    2004   *Blackfoot Ways of Knowing.* Calgary: University of Calgary Press.

Beckwith, Martha Warren
    1937   *Myths and Hunting Stories of the Mandan and Hidatsa Sioux.* New York: AMS Press.

Belyea, Barbara
    1994   *Columbia Journals: David Thompson.* Montreal and Kingston: McGill-Queen's University Press.

Berkes, Fikret
    1993   Traditional Ecological Knowledge in Perspective. In *Traditional Ecological Knowledge: Concepts and Cases,* edited by Julian T. Inglis, p. 1–9. Ottawa: Canadian Museum of Nature/ International Development Research Centre.

Berlin, B.
    1992   *Ethnobiological Classification: Principles for the Categorization of Plants and Animals in Traditional Societies.* Princeton, New Jersey: Princeton University Press.

Bernard, H. Russell
    2006   *Research Methods in Anthropology. Quantitative and Qualitative Approaches.* Oxford, England: AltaMira Press.

Bethke, Brandi, Kacy Hollenback, and Maria Nieves Zedeño
    2014   *Ethnographic Overview and Assessment of Native American Resources in the Niobrara National Scenic River and Lower Missouri National Recreational River.* Report prepared for the National Park Service Midwest Region. Tucson: Bureau of Applied Research in Anthropology, University of Arizona.

Bicker, Alan, Paul Sillitoe, and Johan Pottier
    2004   *Investigating Local Knowledge: New Directions, New Approaches.* Burlington, Vermont: Ashgate.

Biolsi, Thomas, and Larry Zimmerman, editors
    1997   *Indians and Anthropologists: Vine Deloria, Jr., and the Critique of Anthropology.* Tucson: University of Arizona Press.

Black, Mary B.
    1977   Ojibwa Power Belief System. In *Anthropology of Power,* edited by R. Fogelson and R. N. Adams, pp. 141–151. New York: Academic Press.

Blegen, Anne H.
1925   Journal of the Voyage Made by Chevalier de la
        Verendrye, with One of His Brothers, in Search
        of the Western Sea Addressed to the Marquis
        de Beauharnois. *The Quarterly of the Oregon
        Historical Society* 26:116–129.

Blue Talk, Richard
1978   *The Crow: An Assiniboine Story.* The Indian
        Reading Series, Level II Book 13. Portland:
        Northwest Regional Educational Laboratory.

Boster, James
1987   Agreement between Biological Classification
        Systems is not Dependent on Cultural Transmis-
        sion. *American Anthropologist* 89(4):914–920.

Bowers, Alfred W.
1992   [1965] *Hidatsa Social and Ceremonial Organiza-
        tion.* Lincoln: University of Nebraska Press.
2004   [1950] *Mandan Social and Ceremonial Organi-
        zation.* Chicago: University of Chicago Press.

Brink, Jack W., and Robert Dawe
1989   *Final Report of the 1985 and 1986 Field Season
        at Head-Smashed-In Buffalo Jump, Alberta.*
        Edmonton: Alberta Culture and Multicultural-
        ism, Historical Resources Division, Archaeolog-
        ical Survey of Alberta.

Brown, Linda A., and Kitty F. Emery
2008   Negotiations with the Animate Forest: Hunt-
        ing Shrines in the Guatemalan Highlands.
        Journal of Archaeological Method and Theory
        (15)4:300–337.

Brunton, Diane H.
1986   Fatal Antipredator Behavior of a Killdeer. *The
        Wilson Bulletin* 98(4):605–607.

Burpee, Lawrence. J., editor
1907   *York Factory to the Blackfeet Country: The Jour-
        nal of Anthony Henday, 1754–55.* Ottawa: Royal
        Society of Canada.
1909   *Adventurer from Hudson Bay: The Journal of
        Matthew Cocking, from York Factory to the
        Blackfeet Country, 1772–73.* Ottawa: Royal
        Society of Canada.
1927   *Journals and Letters of Pierre Gaultier
        de Varennes de La Verendrye and His Sons.*
        Toronto: Champlain Society.

Catlin, George
1967   *O-kee-pa, A Religious Ceremony, and Other
        Customs of the Mandans.* New Haven: Yale
        University Press.
1989   *North American Indians,* edited by Peter
        Matthiessen. New York: Penguin Group.

Colwell-Chanthaphonh, Chip, and T. J. Ferguson, editors
2008   *Collaboration in Archaeological Practice.* Lan-
        ham, Maryland: Altamira Press.

Chittenden, Hiram M., and Alfred T. Richardson
1905   *Life, Letters, and Travels of Father Pierre-Jean
        De Smet, S.J. 1801–1873,* Vol. 3. New York:
        Francis P. Harper.

Committee on Missouri River Ecosystem Science (CMRES)
2002   *The Missouri River Ecosystem: Exploring the
        Prospects for Recovery.* Washington, D.C.:
        National Academy Press.

Conaty, Jerald, editor
2015   *We Are Coming Home: Repatriation and the
        Restoration of Blackfoot Cultural Confidence.*
        Edmonton: Athabaska University Press.

Coomaraswamy, Ananda K.
2007   *Figures of Speech or Figures of Thought?: The Tra-
        ditional View of Art,* edited by William Wroth.
        Bloomington, Indiana: World Wisdom.

Cornell Laboratory of Ornithology (CLO)
2011   *All About Birds.* Cornell University. http://www
        .allaboutbirds.org, accessed July 8, 2011.

Coues, Elliott
1897   *New Light on the Early History of the Greater
        Northwest: The Manuscript Journals of Alexander
        Henry and David Thompson, 1799–1814,*
        3 Volumes. New York: Francis P. Harper.

Crespi, Muriel
2003   An Overview of the Ethnography Program.
        Paper presented on February 25 at the National
        Park Service TAPS Conference, Tucson, Arizona.

Culin, Steward
1975   *Games of the North American Indians.* New
        York: Dover.

Curtis, Edward S.
1909   *The North American Indian,* Vol. 5: The Mandan.
        The Arikara. The Atsina. Cambridge, Mass.: The
        University Press.
1997   *The North American Indian.* Complete Portfo-
        lios. New York: Taschen.

Davis, Anthony, and John R. Wagner
2003   Who Knows? On the Importance of Identifying
        "Experts" When Researching Local Ecological
        Knowledge. *Human Ecology* 31(3):463–489.

Denig, Edwin Thompson
1961   *Five Indian Tribes of the Upper Missouri: Sioux,
        Arickaras, Assiniboines, Crees, Crows.* Norman:
        University of Oklahoma Press.
1980   *Of the Crow Nation.* New York: AMS Press.
        Reprinted. Originally published 1953, Bureau of

American Ethnology Bulletin No. 151. Washington, D.C.: Smithsonian Institution.

2000 *The Assiniboine.* Edited by J. N. B. Hewitt. Norman: University of Oklahoma Press.

Densmore, Frances
1923 *Mandan and Hidatsa Music.* Bureau of American Ethnology Bulletin No. 80. Washington, D.C.: Smithsonian Institution.

Dollar, Clyde D.
1977 The High Plains Smallpox Epidemic of 1837–38. *The Western Historical Quarterly* 8:15–38.

Dorsey, George A.
1904 *Traditions of the Arikara.* Washington, D.C.: Carnegie Institution.

Dupré, J.
1993 *The Disorder of Things.* Cambridge, Massachusetts.: Harvard University Press.

Dusenberry, Verne
1960 Notes on the Material Culture of the Assiniboine Indians. *Ethnos* 25(1–2):44–62.

Ewers, John C.
1958 *The Blackfeet: Raiders of the Northwestern Plains.* Norman: University of Oklahoma Press.

1974 *Blackfeet Indians: Ethnohistorical Report on the Blackfeet and Gros Ventre Tribes of Indians.* New York: Garland.

1981 The Use of Artifacts and Pictures in the Study of Plains in the Study of Plains Indian History, Art, and Religion. *Annals of the New York Academy of Sciences* 376:247–266.

Fabian, Johannes
1975 Taxonomy and Ideology: On the Boundaries of Concept-classification. In *Linguistics and Anthropology: in Honor of C. F. Voegelin,* edited by Marvin Kinkade, Kenneth Hale and Oswald Werner, pp.183–187. Lisse, Netherlands: Peter de Ridder Press.

Falk, Carl
2002 Fish, Amphibian, Reptile, and Bird Remains. In *Prehistory on First Street NE: The Archaeology of Scattered Village, Mandan, North Dakota,* edited by Stanley Ahler, pp. 7.1–7.25. Submitted to the City of Mandan and North Dakota Department of Transportation. Flagstaff, Arizona: Paleo Cultural Research Group.

Falk, Carl R., Holmes A. Semken, Jr., Lynn M. Snyder, Carole A. Angus, Darcy F. Morey, Danny E. Olinger, and R. S. Rosenberg
1991 *Inventory of Identified Vertebrate Specimens from Phase I Archeological Investigations at the Knife River Indian Villages National Historic Site, Mercer County, North Dakota.* Department of Anthropology Contribution No. 267. Grand Forks: University of North Dakota.

Federal Register
1940 Bald Eagle Protection Act, 16 U.S.C. 668–668d, 54 Stat. 250; Presidential Memorandum 59 F.R. 22953).

Feld, S.
1990 *Sound and Sentiment: Birds, Weeping, Poetics, and Song in Kaluli Expressions.* Philadelphia: University of Pennsylvania Press.

Fitzgerald, Michael O.
1991 *Yellowtail, Crow Medicine Man and Sun Dance Chief: An Autobiography.* Norman: University of Oklahoma Press.

Fort Belknap Indian Community
2003 Fort Belknap Indian Community Official Website. http://www.ftbelknap.org, accessed March 2012.

Fort Peck Assiniboine & Sioux Tribes
2011 Fort Peck Assiniboine & Sioux Tribes. http://www.fortpecktribes.org, accessed March 2012.

Frey, Rodney
1987 *The World of the Crow Indians: As Driftwood Lodges.* Norman: University of Oklahoma Press.

Frison, George C.
1979 The Crow Indian Occupation of the High Plains: The Archaeological Evidence. *Archaeology in Montana* 203:3–16.

Garcia, Louis
1995 Hidatsa Place Names. Unpublished MS, Mandan, Hidatsa and Arikara Cultural Preservation Office, New Town, North Dakota.

Gell, Anthony
1998 *Art and Agency: An Anthropological Theory.* New York: Clarendon Press.

Gibbon, Guy
2003 *The Sioux: Dakota and Lakota Nations.* Oxford: Blackwell.

Gilman, Carolyn, and Mary Jane Schneider
1987 *The Way to Independence: Memories of a Hidatsa Indian Family, 1840–1920.* St. Paul: Minnesota Historical Society Press.

Gilmore, Melvin Randolph
1929 *Arikara Account of the Origin of Tobacco and Catching of Eagles.* Indian Notes No. 6. New York: Museum of the American Indian, Heye Foundation.

Gilmore, Melvin Randolph (*continued*)
1930 *The Arikara Book of Genesis.* Papers of the Michigan Academy of Science, Arts, and Letters 12:95–120.
1932 *The Sacred Bundles of the Arikara.* Ann Arbor: University of Michigan. Reprint from Papers of the Michigan Academy of Science, Arts, and Letters 16:33–50.

Gosler, Andrew, Deborah Buehler, and Alberto Castillo
2010 The Broader Significance of Ethno-ornithology. In *Ethno-ornithology: Birds, Indigenous Peoples, Culture, and Society,* edited by Sonia Tidemann and Andrew Gosler, pp. 31–46. Washington, D.C.: Earthscan.

Gough, Barry M., and Robert Craig Brown
1988 *The Journal of Alexander Henry the Younger 1799–1814.* Toronto: Champlain Society.

Greiser, Sally T.
1994 Late Prehistoric Cultures in the Montana Plains. In *Plains Indians,* A.D. *500–1500, The Archaeological Past of Historic Groups,* edited by K. H. Schlesier, pp. 34–55. Norman: University of Oklahoma.

Grinnell, George Bird
1913 *Blackfeet Indian Stories.* New York: Charles Scribner's Sons.
1920 *Blackfoot Lodge Tales: The Story of a Prairie People.* New York: Charles Scribner's Sons.

Hallowell, Irving
1976 Ojibwa Ontology, Behavior, and Worldview. In *Irving Hallowell Contributions to Anthropology,* pp. 357–390. Chicago: University of Chicago Press.

Hanson, Jeffery R.
1980 Structure and Complexity of Medicine Bundle Systems of Selected Plains Tribes. *Plains Anthropologist* 25(89):199–216.

Hill, Matthew G., and D. J. Rapson
2008 Unresolved Taphonomic Histories, Interpretive Equivalence, and Paleoindian Faunal Exploitation. Paper presented at the 73rd annual meeting of the Society of American Archaeology, Vancouver, British Columbia.

Hodder, Ian
2012 *Entangled: An Archaeology of the Relationships between People and Things.* Malden, Massachusetts: John Wiley & Sons.

Hollenback, Kacy L.
2010 Social Memory of Disaster: Exploring Historic Smallpox Epidemics among the Mandan and Hidatsa. Paper presented at the 68th annual meeting of the Plains Anthropological Society, Bismarck, North Dakota.
2012 *Disaster, Technology, and Community: Measuring Responses to Smallpox Epidemics in Historic Hidatsa Villages, North Dakota.* Ph.D. Dissertation, School of Anthropology, University of Arizona, Tucson.

Hungry-Wolf, Adolf
2006 *The Blackfoot Papers,* vol. I: Pikunni History and Culture. Suokumchuck, British Columbia: The Good Medicine Cultural Foundation.

Hudson Bay Company Archives (HBCA)
n.d. York Factory Fur Returns, B239/1. Winnipeg: Provincial Archives of Manitoba.

Hunt, David, Joseph C. Porter, and W. Raymond Wood
2002 *Karl Bodmer's Studio Art. The Newberry Library Bodmer Collection.* Chicago: University of Illinois Board of Trustees.

Ingold, Tim
2000 *The Perception of the Environment. Essays on Livelihood, Dwelling and Skill.* London: Routledge.

Irwin, Lee
1994 *The Dream Seekers: Native American Visionary Traditions of the Great Plains.* Norman: University of Oklahoma Press.

Jackson, John C.
2000 *The Piikani Blackfeet.* Missoula, Montana: Mountain Press.

Jenkinson, Clay, and James P. Ronda
2003 *A Vast and Open Plain: The Writings of the Lewis and Clark Expedition in North Dakota 1806–1808.* Bismarck: State Historical Society of North Dakota.

Johnsgard, Paul A.
2012 *Wings Over the Great Plains: Bird Migrations in the Central Flyway.* Zea E-Books, book 13. http://digitalcommons.unl.edu/zeabook/13. Accessed May 19, 2016.

Johnson, Craig M.
2007 *A Chronology of Middle Missouri Plains Village Sites.* Contributions to Anthropology No. 47. Washington, D.C.: Smithsonian Institution Scholarly Press.

Keane, Webb
2003 Semiotics and the Social Analysis of Material Things. *Language & Communication* 23:409–425.

Kennedy, Michael Stephen, editor
1961 *The Assiniboines.* Norman: University of Oklahoma Press.

Krech, Shepard, III
2009   *Spirits of the Air: Birds and American Indians in the South.* Athens: University of Georgia Press.

Kvamme, Kenneth L., and Stanley A. Ahler
2007   Integrated Remote Sensing and Excavation at the Double Ditch State Historic Site, North Dakota. *American Antiquity* 72(3):539–561.

Larpenteur, Charles
1898   *Forty Years a Fur Trader on the Upper Missouri,* edited by Elliott Coues. New York: F. P. Harper.

Laustrup, Mark, and Mike LeValley
1998   , *Missouri River Environmental Assessment Program.* Columbia, Missouri: U.S. Geological Survey.

Lawson, Michael L.
1982   *Dammed Indians: The Pick-Sloan Plan and the Missouri River Sioux, 1944–1980.* Norman: University of Oklahoma Press.
1994   *Dammed Indians Revisited: The Continuing History of the Pick-Sloan Plan and the Missouri River Sioux.* Pierre: South Dakota State Historical Society Press.

Linderman, Frank B.
1972   *Pretty Shield: Medicine Woman of the Crows.* Lincoln: University of Nebraska Press.
2002   *Plenty-Coups: Chief of the Crows.* Lincoln: University of Nebraska Press.

Lohmann, Roger Ivar
2003   The Supernatural is Everywhere: Defining Qualities of Religion in Melanesia and Beyond. *Anthropological Forum* 13(2):175–185.

Lokensgard, Kenneth Hayes
2010   *Blackfoot Religion and the Consequences of Cultural Commoditization.* Burlington, Vermont: Ashgate.

Lowie, Robert H.
1909   *The Assiniboine.* Anthropological Papers No. 4(1):1–271. New York: American Museum of Natural History.
1915a   *Societies of the Arikara Indians.* Anthropological Papers No. 11(8):647–678. New York: American Museum of Natural History.
1915b   *The Sun Dance of the Crow Indians.* Anthropological Papers No. 16(1):1–50. New York: American Museum of Natural History.
1918   *Myths and Traditions of the Crow Indians.* Anthropological Papers No. 25(1):3–308. New York: American Museum of Natural History.
1919   *The Tobacco Society of the Crow Indians.* Anthropological Papers No. 21(2):101–200. New York: American Museum of Natural History.
1922a   *The Material Culture of the Crow Indians.* Anthropological Papers No. 21(3):201–270. New York: American Museum of Natural History.
1922b   *The Religion of the Crow Indians.* Anthropological Papers No. 25(2):309–444. New York: American Museum of Natural History.
1924   *Minor Ceremonies of the Crow Indians.* Anthropological Papers No. 21(5):323–365. New York: American Museum of Natural History.
1939   *Hidatsa Texts.* Indianapolis: Indiana Historical Society.
1954   Indians of the Plains. In *Anthropological Handbook* No. 1. Published for the American Museum of Natural History. New York: McGraw-Hill.
1956   *The Crow Indians.* New York: Rinehard.
1960   A Few Assiniboine Texts. *Anthropological Linguistics* 2(8): 1–30.

Mails, Thomas E.
1973   *Soldiers, Bear Men, and Buffalo Women: A Study of the Societies and Cults of the Plains Indians.* Englewood Cliffs: Prentice-Hall.

March, H. Colley
1898   Mythology of Wise Birds. *The Journal of the Anthropological Institute of Great Britain and Ireland* 27:209–232.

Matthews, Washington
1877   *Ethnography and Philology of the Hidatsa Indians.* Washington, D.C.: U.S. Government Printing Office.

McCleary, Timothy P.
1997   *The Stars We Know: Crow Indian Astronomy and Lifeways.* Prospect Heights, Illinois: Waveland Press.

McClintock, Walter
1999   [1910] *The Old North Trail: Or, the Life, Legends, and Religion of the Blackfeet Indians.* London: MacMillan.

McGowan, Ernest. S.
1942   The Arikara Indians. *Minnesota Archaeologist* 8:83–122.

Medicine Crow, Joe
2007   Introduction. In *Native Spirit: The Sun Dance Way,* as told by Thomas Yellowtail, p. 1. Bloomington, Indiana: World Wisdom.
n.d.   Interview with Joe Medicine Crow. (http://www.custermuseum.org/ medicinecrow.htm

Meyer, Carter Jones, and Diana Royer, editors
2001   *Selling the Indian: Commercializing and Appropriating American Indian Cultures.* Tucson: University of Arizona Press.

Mills, Barbara, and T. J. Ferguson
    2008   Animate Objects: Shell Trumpets and Ritual Networks in the Greater Southwest. *Journal of Archaeological Method and Theory* 15:338–361.

Mills, Barbara J., and William H. Walker, editors
    2008   *Memory Work: Archaeologies of Material Practice.* Santa Fe: SAR Press.

Mitchell, Mark
    2013   *Crafting History in the Northern Plains: A Political Economy of the Heart River Region 1400–1750.* Tucson: University of Arizona Press.

Montana Writer's Project (MWP)
    1942   *Land of Nakoda: The Story of the Assiniboine Indians.* Helena: State Publishing Company.

Moore, John H.
    1986   The Ornithology of Cheyenne Religionists. *Plains Anthropologist* 31(113):177–192.

Moore, Kaytlin
    2012   Negotiating the Middle Ground in a World-System: The Niitsitapi (Blackfoot) and Ktunaxa (Kootenai) in the Northern Rocky Mountain Fur Trade. Master's Report, School of Anthropology, University of Arizona, Tucson.

Murray, Wendi Field
    2009   "The Gods Above Have Come": A Contemporary Analysis of the Eagle as a Cultural Resource in the Northern Plains. Master's thesis, School of Anthropology, University of Arizona, Tucson.
    2011   Feathers, Fasting, and the Eagle Complex: A Contemporary Analysis of the Eagle as a Cultural Resource in the Northern Plains. *Plains Anthropologist* 56(218):143–154.

Murray, Wendi Field, and Fern Swenson
    2016   Situational Sedentism. In *Arikara Yesterday and Today,* edited by Kacy Hollenback. Plains Anthropologist Memoir. In press.

Murray, Wendi Field, María Nieves Zedeño, Kacy L. Hollenback, Calvin Grinnell, and Elgin Crows Breast
    2011   The Remaking of Lake Sakakawea: Locating Cultural Viability in Negative Heritage on the Missouri River. *American Ethnologist* 38(3):468–483.

Nabokov, Peter
    1970   *Two Leggings: The Making of a Crow Warrior.* Apollo Edition. New York: Thomas Y. Crowell.

Nadasdy, Paul
    2007   The Gift in the Animal: The Ontology of Hunting and Human-Animal Sociality. *American Ethnologist* 34(1):28–43.

National Anthropological Archives
    1929   Manuscript 1929-a. Department of Anthropology, National Museum of Natural History, Smithsonian Institution, Washington, D.C.

Northern Prairie Wildlife Research Center (NPWRC)
    2006   Range Expansion of the Pileated Woodpecker in North Dakota. http://www.npwrc.usgs.gov/resource/birds/pwprange/observ.htm, accessed September 26, 2011.

Parks, Douglas R.
    1986   *An English-Arikara Student Dictionary.* Roseglen, North Dakota: White Shield School District.
    1991   *Traditional Narratives of the Arikara Indians.* Lincoln: University of Nebraska Press, in cooperation with the American Indian Studies Research Institute, Indiana University.
    1996   *Myths and Traditions of the Arikara Indians.* Lincoln: University of Nebraska Press.
    2001   Arikara. In *Plains,* part 1, edited by Raymond J. DeMallie, pp. 365—390. Handbook of North American Indians, vol.13, William C. Sturtevant, general editor. Washington, D.C.: Smithsonian Institution.

Parks, Douglas R., A. Wesley Jones, and Robert C. Hollow, editors
    1978   *Earth Lodge Tales from the Upper Missouri: Traditional Stories of the Arikara, Hidatsa, and Mandan.* Bismarck, North Dakota: University of Mary.

Parmalee, Paul W.
    1977a   Avian Bone Pathologies from Arikara Sites in South Dakota. *The Wilson Bulletin* 89(4):628–632.
    1977b   Avifauna from Prehistoric Arikara Sites in South Dakota. *The Plains Anthropologist* 22(77):189–222.
    1979   Inferred Arikara Subsistence Patterns Based on a Selected Faunal Assemblage from Mobridge Site, South Dakota. *Kiva* 44(2/3):191–218.
    1980   Bird Remains from the Bagnell Site (32OL16), Oliver County, North Dakota. *South Dakota Bird Notes* 32(4):75–77.

Pauketat, Timothy
    2012   *An Archaeology of the Cosmos: Rethinking Agency and Religion in Ancient America.* London: Routledge.

Peck, Trevor
    2011   *Light from Ancient Campfires: Archaeological Evidence for Native Lifeways on the Northern*

Plains. Edmonton, Alberta: Athabaska University Press.

Perrins, Christopher M., editor
2009   *The Encyclopedia of Birds*. Oxford: New York.

Peters, Virginia Bergman
1995   *Women of the Earthlodges: Tribal Life in the Plains*. New Haven: Archon Books.

Posey, Darrell
2004   *Indigenous Knowledge and Ethics*. New York: Routledge.

Prucha, Francis Paul
1995   *The Great Father: The United States Government and the American Indians*. Lincoln: University of Nebraska Press.

Prummel, W., J. T. Zeiler, and D. C. Brinkhuizen, editors
2010   *Birds in Archaeology*. Proceedings of the 6th Meeting of the ICAZ Bird Working Group in Groningen. Groningen, Netherlands: Stitching Groningen Universiteitsfonds.

Ray, Arthur J.
1974   *Indians in the Fur Trade: Their Role as Trappers, Hunters, and Middlemen in the Lands Southwest of Hudson Bay, 1660–1870*. Toronto and Buffalo: University of Toronto Press.

Reed, George
n.d.   *Apsaalooke or Crow Tribe*. Crow Agency, Montana: Apsaaloke Nation Cultural Affairs Department.

Reeves, Brian O.K.
1983   *Culture Changes in the Northern Plains 1000 B.C.-A.D. 1000*. Edmonton: Alberta Culture Historical Resources Division.
1993   Iniskim: A Sacred Piikáni Religious Tradition. In *Kunaitapii: Coming Together on Native Sacred Sites*, edited by B.O.K. Reeves and M. Kennedy, pp. 194–247. Calgary: Archaeological Society of Alberta.

Reeves, Brian O.K., and Sandra Peacock
2001   Our Mountains Are Our Pillows: An Ethnographic Overview of Glacier National Park. MS on file at the Blackfeet Tribal Historic Preservation Office, Browning, Montana.

Rhodes, Richard
2006   *John James Audubon: The Making of an American*. New York: Knopf.

Richert, Bernhard. E.
1969   Plains Indian Medicine Bundles. Master's Thesis, Department of Anthropology, University of Texas, Austin.

Rodnick, David
1938   *Fort Belknap Assiniboine of Montana a Study in Culture Change*. Privately published by the author. New Haven, Connecticut. Available from Hathi Trust Digital Library.

Rosaldo, Michelle Z.
1972   Metaphors and Folk Classification. *Southwestern Journal of Anthropology* 28(1):83–99.

Schaeffer, Claude E.
1950   *Bird Nomenclature and Principles of Avian Taxonomy of the Blackfeet Indians*. Washington, D.C: Washington Academy of Sciences.

Schaeffer, Claude E., and Mrs. Schaeffer
1934   Blackfoot Papers. Calgary, Canada: Glenbow Museum Archives.

Schneider, Mary Jane
2001   Three Affiliated Tribes. In *Plains,* edited by Ray J. DeMallie, pp. 391–398. Handbook of North American Indians, vol. 13, William C. Sturtevant, general editor. Washington D.C: Smithsonian Institution Press.
2004   Native American Traditional Art. In *Encyclopedia of the Great Plains,* edited by David Wishart, p. 125. Lincoln: University of Nebraska Press.

Schultz, James Willard
1916   *Blackfeet Tales of Glacier National Park*. Boston and New York: Houghton Mifflin.
1962   *Blackfeet and Buffalo*. Norman: University of Oklahoma Press.

Scriver, Bob
1990   *Blackfeet: Artists of the Northern Plains: The Scriver Collection of Blackfeet Indian Artifacts and Related Objects, 1894–1990*. Kansas City, Missouri: Lowell Press.

Sillitoe, Paul, Alan Bicker, and Johan Pottier, editors
2002   *Participating in Development: Approaches to Indigenous Knowledge*. New York: Routledge.

Sterry, Paul, and Brian E. Small
2009   *Birds of Western North America: A Photographic Guide*. Princeton and Oxford: Princeton University Press.

Stoffle, Richard W., Maria Nieves Zedeño, and David B. Halmo, editors
2001   *American Indian and the Nevada Test Site: A Model of Research and Consultation*. Tucson: Bureau of Applied Research in Anthropology, University of Arizona.

Strong, William Duncan
1933   *Studying the Arikara and the Neighbors on the Upper Missouri, Explorations and Field-Work of*

Strong, William Duncan (*continued*)
　　Smithsonian Institution in 1932. Publication
　　No. 3213, pp. 73–76. Washington, D.C.: Smith-
　　sonian Institution.

Thwaites, Reuben Gold, editor
　1904　*Original Journals of the Lewis and Clark Expedi-
　　tion: 1804–1806,* Vol. 1. New York: Dodd, Mead.
　1905a　*Early Western Travels 1748–1846,* Vol. 22. Cleve-
　　land: Arthur H. Clark.
　1905b　*Original Journals of the Lewis and Clark Expedi-
　　tion: 1804–1806,* Vol. 3. New York: Dodd, Mead.
　1906　*Early Western Travels 1748–1846,* Vol. 23. Cleve-
　　land: Arthur H. Clark.

Tidemann, Sonia, and Andrew Gosler, editors
　2010　*Ethno-ornithology: Birds, Indigenous Peo-
　　ples, Culture, and Society.* Washington, D.C.:
　　Earthscan.

Tidemann, Sonia, and Tim Whiteside
　2010　Aboriginal Stories: The Riches and Colour of
　　Australian Birds. In *Ethno-ornithology: Birds,
　　Indigenous Peoples, Culture, and Society,* edited
　　by Sonia Tidemann and Andrew Gosler,
　　pp. 153–179. Washington, D.C.: Earthscan.

Tyrrell, J. B., editor
　1931　*Documents Relating to the Early History of Hud-
　　son Bay.* Toronto: The Champlain Society.
　1934　*Journals of Samuel Hearne and Philip Turnor.*
　　Publications Vol. 21. Toronto: Champlain
　　Society.

Ubelaker, Douglas H., and Waldo R. Wedel
　1975　Bird Bones, Burials and Bundles in Plains
　　Archaeology. *American Antiquity* 40(4):444–452.

VanStone, James W.
　1996　*Ethnographic Collections from the Assiniboine
　　and Yanktonai Sioux in the Field Museum of
　　Natural History.* Chicago: Field Museum of
　　Natural History.

Viveiros de Castro, Edward
　2004　Exchanging Perspectives: The Transformation
　　of Objects into Subjects in Amerindian Ontol-
　　ogies. *Common Knowledge* 10:463–485.

Voget, Fred W.
　1995　*They Call Me Agnes: A Crow Narrative Based
　　on the Life of Agnes Yellowtail Deernose,* assisted
　　by Mary K. Mee. Norman: University of Okla-
　　homa Press.
　2001　Crow. In *Plains,* edited by Ray J. DeMallie,
　　pp. 695–717. Handbook of North American
　　Indians, Vol 13, William C. Sturtevant, general
　　editor. Washington, D.C.: Smithsonian Institu-
　　tion Press.

Walde, Dale
　2006　Sedentism and Pre-contact Tribal Organization
　　on the Northern Plains: Colonial Imposition or
　　Indigenous Development? *World Archaeology*
　　38(2):291–310.

Wallaert, Helene
　2006　Beads and a Vision: Waking Dreams and
　　Induced Dreams as a Source of Knowledge for
　　Beadwork Making. An Ethnographic Account
　　from Sioux Country. *Plains Anthropologist*
　　51(197):3–15.

Watts, Christopher
　2013　*Relational Archaeologies: Humans, Animals,
　　Things.* New York: Routledge.

Wavey, Robert
　1993　International Workshop on Indigenous Knowl-
　　edge and Community Based Resource Manage-
　　ment: Keynote Address. In *Traditional Ecological
　　Knowledge: Concepts and Cases,* edited by J. T.
　　Inglis, pp. 11–16. Ottawa: International Devel-
　　opment Research Centre.

Weitzner, Bella
　1979　*Notes on the Hidatsa Indians Based on Data
　　Recorded by the Late Gilbert L. Wilson.* Anthro-
　　pological Papers No. 56(2). New York: American
　　Museum of Natural History.

Western Heritage Center (WHC)
　2008　Crow Expressions. Online exhibit, http://www
　　.ywhc.org/index.php?p=83, accessed Novem-
　　ber 17, 2011.

Wildschut, William
　1975　*Crow Indian Medicine Bundles.* New York:
　　Museum of the American Indian, Heye
　　Foundation.

Will, George F.
　1930　Arikara Ceremonials. *North Dakota Historical
　　Quarterly* 4(4):247–265.
　1934　*Notes on the Arikara Indians and their Ceremo-
　　nies.* Denver: John VanMale.

Will, George F., and George E. Hyde
　1964　*Corn among the Indians of the Upper Missouri.*
　　Lincoln: University of Nebraska Press.

Will, George F., and Herbert J. Spinden
　1906　*The Mandans: A Study of their Culture, Archae-
　　ology, and Language.* Cambridge, Mass.: Pea-
　　body Museum.

Wilson, Gilbert L.
　1908　Field Notes. On file at the Minnesota State
　　Historical Society, St. Paul.
　1909　Hidatsa-Mandan Report. Ms. on file at the
　　Minnesota State Historical Society, St. Paul.

1910 Butterfly and Wounded Face, from field notes. Ms. on file at the Minnesota State Historical Society, St. Paul.

1911 Field Notes. Ms. on file at the Minnesota State Historical Society, St. Paul.

1924 *The Horse and Dog in Hidatsa Culture.* Anthropological Papers No. 15(2). New York: American Museum of Natural History.

1928 Hidatsa Eagle Trapping. Anthropological Papers No. 30(4):99–245. New York: American Museum of Natural History.

Wissler, Clark
1912 *Social Organization and Ritualistic Ceremonies of the Blackfoot Indians.* Anthropological Papers No. 7. New York: American Museum of Natural History.

Wissler, Clark, and D. C. Duvall
1908 *Mythology of the Blackfoot Indians.* Anthropological Papers No. 2(1):1–164. New York: American Museum of Natural History.

Witte, Stephen S., and Marsha V. Gallagher, editors
2008 *The North American Journals of Prince Maximilian of Wied,* vol. 2, April-September 1833. Norman: University of Oklahoma Press.

Wood, Raymond W.
1986 *The Origins of the Hidatsa Indians: A Review of the Ethnohistorical and Traditional Data.* Reprints in Anthropology, Vol. 32. Lincoln: J & L Reprint Company.

Wood, W. Raymond, and Allan S. Downer
1977 *Notes on the Crow-Hidatsa Schism. Trends in Middle Missouri Prehistory.* Memoir No. 13(22):83–100. Lincoln: Plains Anthropology.

Wood, W. Raymond, and Thomas D. Thiessen, editors
1985 *Early Fur Trade on the Northern Plains: Canadian Traders Among the Mandan and Hidatsa Indians, 1738–1818.* Norman: University of Oklahoma Press.

Zarrillo, Sonia, and Brian P. Kooyman
2006 Evidence for Berry and Maize Processing on the Canadian Plains from Starch Grain Analysis. *American Antiquity* 71:473–499.

Zedeño, Maria Nieves
2008a Bundled Worlds: The Roles and Interactions of Complex Objects from the North American Plains. *Journal of Archaeological Method and Theory* 15:362–378.

2008b Traditional Knowledge, Ritual Behavior, and Contemporary Interpretation of the Archaeological Record. In *Belief in the Past,* edited by Kelly Hayes-Gilpin and David Whitley, pp. 259–274. Walnut Creek, California: Left Coast Press.

2009 Animating by Association: Index Objects and Relational Taxonomies. *Cambridge Archaeological Journal* 19(3):407–417.

2013 Methodological and Analytical Challenges in Relational Archaeologies, a View from the Hunting Ground. In *Relational Archaeologies: Humans, Animals, Things,* edited by Christopher Watts, pp. 117–134. New York: Routledge.

Zedeño, María Nieves, Jesse A. Ballenger, and John R. Murray
2014 Landscape Engineering and Organizational Complexity Among Late Prehistoric Bison Hunters of the Northwestern Plains. *Current Anthropology* 55:23–58.

Zedeño, Maria Nieves, Christopher Basaldú, and Amy Eisenberg
2001 *Cultural Affiliation and Ethnographic Assessment of the St. Croix National Recreational River, Wisconsin and Minnesota.* Report prepared for the National Park Service Midwest Region. Tucson: Bureau of Applied Research in Anthropology, University of Arizona.

Zedeño, María Nieves, Kacy Hollenback, Christopher Basaldú, Vania Fletcher, and Samrat Miller
2006 *Cultural Affiliation Statement and Ethnographic Resource Assessment Study for Knife River Indian Villages N.H.S., Fort Union Trading Post N.H.S., and Theodore Roosevelt National Park, North Dakota.* Prepared for the National Park Service Midwest Regional Office. Tucson: Bureau of Applied Research in Anthropology, University of Arizona.

Zedeño, Maria Nieves, Kacy Hollenback, and Calvin Grinnell
2009 From Path to Myth: Journeys and the Naturalization of Nation along the Upper Missouri River. In *The Anthropology of Paths and Trails,* edited by James Snead, J. Andrew Darling, and Carl L. Erickson, pp. 106–133. Philadelphia: University of Pennsylvania Press.

Zedeño, María Nieves, Wendi F. Murray, and John R. Murray
2015 Central Places in the Back Country: The Archaeology and Ethnography of Beaver Lake, Montana. In *Engineering Mountain Landscapes: An Archaeology of Social Investment,* edited by Laura Scheiber and María Nieves Zedeño, pp. 7–22. Salt Lake City: University of Utah Press.

Index

ABSTRACT

The Missouri River Basin is home to thousands of bird species that migrate across the Great Plains of North America each year, marking the seasonal cycle and filling the air with their song. Since time immemorial, Native inhabitants of this vast region established alliances with birds that helped them to connect with the gods, to learn the workings of nature, and to live well. This ethno-ornithology integrates published and archival sources covering archaeology, ethnohistory, historical ethnography, folklore, and interviews with contemporary Native American elders from six ethnic groups—Blackfoot, Assiniboine, Mandan, Hidatsa, Arikara, and Crow—to characterize the society of people and birds. These sources further provide a deep understanding of how birds are situated in contemporary practice, and what has fostered the cultural persistence of human-bird relationships that began with the creation of the world.

Native principles of ecological and cosmological knowledge are brought into focus to highlight specific beliefs, practices, and concerns associated with individual bird species, bird parts, bird objects, and the natural and cultural landscapes that birds and people cohabit. The world of bird-human relationships is laid out in seven major themes, beginning with indigenous notions of the constitution of birds and continuing with their role in the creation of the world, their unique personalities, behaviors, abilities, and communicative skills. The materiality of birds' extraordinary alliance with humans is unpacked through imagery and objects, and their past and present significance in realms of daily life such as subsistence, trade, ritual, spirituality, politics, and war is underscored. The book concludes with a reflection on the future of the society of people and birds.

RESUMEN

La Cuenca del Río Missouri es un repositorio de miles de especies de aves que emigran anualmente a través los Grandes Llanos de Norte América, demarcando el ciclo de estaciones y llenando el aire con su canto. En tiempo inmemorial, los indígenas the esta región establecieron alianzas con la aves, quienes les ayudaron a conectarse con sus dioses, a familiarizarse con la naturaleza, y a vivir bien. Esta etno-ornitología integra publicaciones y archivos abarcando arqueología, ethnohistoria, ethnografía histórica, folklore, y entrevistas con miembros de seis grupos étnicos—Blackfoot, Assiniboine, Mandan, Hidatsa, Arikara y Crow—para así caracterizar la sociedad the aves y gente. Estas fuentes además proveen un entendimiento profundo sobre el lugar que las aves ocupan en la práctica cultural contemporánea y los factores que han estimulado la persistencia de las relaciones entre aves y humanos.

Se emplean principios y conocimiento indígenas de ecología y cosmología para realzar creencias, actividades, y preocupaciones asociadas con ciertas especies de aves y con sus partes, con cultura material, y con paisajes naturales y culturales compartidos por la aves y la gente. Este mundo de relaciones se desarrolla en siete temas, empezando con la constitución de las aves, sus personalidades, hábitos, y habilidades, y sus poderes de comunicación. Se estudia la materialidad de esta extraordinaria alianza a través de imágenes y objetos y su significado en subsistencia, intercambio, ritual, espiritualidad, política, y guerra. El libro concluye con una reflección acerca de el futuro de la sociedad de aves y gente.

ANTHROPOLOGICAL PAPERS OF THE UNIVERSITY OF ARIZONA